Robert Steinbauer

Regulation of ribosome biogenesis and RNA polymerase I transcription

Robert Steinbauer

Regulation of ribosome biogenesis and RNA polymerase I transcription

How nutrients control the synthesis of ribosomes

Südwestdeutscher Verlag für Hochschulschriften

Imprint

Any brand names and product names mentioned in this book are subject to trademark, brand or patent protection and are trademarks or registered trademarks of their respective holders. The use of brand names, product names, common names, trade names, product descriptions etc. even without a particular marking in this work is in no way to be construed to mean that such names may be regarded as unrestricted in respect of trademark and brand protection legislation and could thus be used by anyone.

Publisher:
Südwestdeutscher Verlag für Hochschulschriften
is a trademark of
Dodo Books Indian Ocean Ltd., member of the OmniScriptum S.R.L Publishing group
str. A.Russo 15, of. 61, Chisinau-2068, Republic of Moldova Europe
Printed at: see last page
ISBN: 978-3-8381-2614-2

Zugl. / Approved by: Regensburg, University, Diss., 2011

Copyright © Robert Steinbauer
Copyright © 2011 Dodo Books Indian Ocean Ltd., member of the OmniScriptum S.R.L Publishing group

Table of Contents

1 SUMMARY 1
    Summary 1

2 INTRODUCTION 3
    2.1   Ribosome biogenesis and cell growth 3
    2.2   RNA polymerase I transcription in eukaryotes 4
        2.2.1   Structure of ribosomal RNA genes 4
        2.2.2   Subunit composition of RNA polymerase I 5
        2.2.3   RNA polymerase I transcription factors and their function 7
        2.2.4   Pre-rRNA processing and rRNA maturation 10
        2.2.5   Pol5p and Mybbp1a – potential regulators of ribosomal RNA synthesis 13
    2.3   TOR – a central component of the eukaryotic growth regulatory network 14
        2.3.1   General description of the target of rapamycin (TOR) 14
        2.3.2   Upstream and downstream of the TOR signaling network 16
        2.3.3   TOR signaling in the context of growth-dependent regulation of ribosome biogenesis 19
    2.4   Objectives 22

3 RESULTS 25
    3.1   Effects of TOR inactivation on RNA polymerase I transcription, rRNA production, and yeast cell growth 25
        3.1.1   Proteasome-dependent reduction of Rrn3p-levels in growth-arrested yeast cells . 25
        3.1.2   Level of Rrn3p influences Pol I-Rrn3p complex formation, Pol I recruitment to the rDNA, and yeast cell growth but not the rDNA copy number 28
        3.1.3   RNA polymerase I transcription is not affected at early stages of TOR inactivation in yeast cells 34
        3.1.4   Inhibition of translation is sufficient to mimic severe pre-rRNA processing defects observed at early stages of TOR inactivation in yeast cells 40
        3.1.5   Short-term TOR inactivation predominantly affects expression of ribosomal proteins whose abundance is important for yeast cell growth 44

	3.1.6	Nucleolar entrapment of ribosome biogenesis factors in yeast cells is mediated by both rapamycin and cycloheximide treatment as well as by conditional shut-down of ribosomal protein expression 48
3.2		**Effects of overexpression of Rrn3p on RNA polymerase I transcription** **51**
	3.2.1	*GAL1*-dependent overexpression of Rrn3p results in defects of yeast cell growth .. 51
	3.2.2	Overexpression of Rrn3p leads to increased amounts of Pol I-Rrn3p complexes in yeast cells.. 53
	3.2.3	ChIP experiments reveal no increase in the association of Pol I with the rDNA locus, but an enhanced level of Rrn3p crosslinking to the rDNA locus when Rrn3p is overexpressed in yeast cells 55
	3.2.4	Overexpression of Rrn3p does not lead to severe pre-rRNA processing defects or changes in mature rRNA production in yeast cells... 58
3.3		**Pol5p, which plays an important role in rRNA synthesis, is a putative interaction partner of Rrn3p** **60**
	3.3.1	Co-purification of Pol5p in the course of phosphorylation analyses of Rrn3p indicates interaction between the two proteins 60
	3.3.2	ChIP experiments reveal no association of Pol5p with the rDNA locus 63

4 DISCUSSION 65

4.1	The role of the proteasome in the down-regulation of Rrn3p-levels upon TOR inactivation... 65
4.2.	The role of Rrn3p-levels in the formation of Pol I-Rrn3p complexes upon TOR inactivation... 66
4.3.	The role of phosphorylation in the formation of Pol I-Rrn3p complexes................. 67
4.4	Uncoupling RNA polymerase I transcription and mature rRNA production after short-term TOR inactivation... 69
4.5	A model for the drastic down-regulation of ribosome production upon TOR inactivation... 71
4.6	Overexpression of Rrn3p and its impact on ribosome biogenesis and yeast cell growth 72
4.7	The role of Pol5p in ribosome biogenesis and yeast cell growth............................... 74
4.8	Outlook 75

5 MATERIAL AND METHODS 77

5.1	Material 77

5.1.1	*Saccharomyces cerevisiae* strains	77
5.1.2	*Escherichia coli* strains	82
5.1.3	Plasmids	82
5.1.4	Oligonucleotides	84
5.1.5	Probes	87
5.1.6	Antibodies	87
5.1.7	Enzymes	88
5.1.8	Kits	88
5.1.9	Media	88
5.1.10	Buffers	91
5.1.11	Chemicals	93
5.1.12	Other materials	94
5.1.13	Equipment	94
5.1.14	Software	95
5.2	**Methods**	**96**
5.2.1	Work with *Saccharomyces cerevisiae*	96
5.2.2	Work with *Escherichia coli*	98
5.2.3	Work with DNA	99
5.2.4	Work with RNA	103
5.2.5	Work with proteins	104
5.2.6	Additional biochemical methods	108

6 REFERENCES **115**

7 PUBLICATIONS **133**

8 ABBREVIATIONS **135**

Acknowledgments **137**

Table of Contents

1 SUMMARY

Summary

Eukaryotic cell growth is tightly linked to the synthesis of new ribosomes, the molecular machineries responsible for protein production. The transcription of a ribosomal precursor RNA (pre-rRNA) by RNA polymerase I (Pol I) constitutes an initial and central step in the complex process of ribosome biogenesis and is therefore one of the main targets for regulation. The initiation of each round of transcription is dependent on the formation of a complex between Pol I and the essential transcription factor Rrn3p. Subsequent processing of this precursor transcript yields three of the four mature ribosomal RNAs (rRNAs) forming a scaffold to which ribosomal proteins (r-proteins/RPs) assemble in the course of ribosome maturation.

Since ribosome biogenesis is one of the most energy-consuming cellular processes, eukaryotic cells cease the production of ribosomes very rapidly upon unfavorable growth conditions like nutrient deprivation in order to ensure survival. The conserved target of rapamycin (TOR)-pathway plays an essential role in both sensing environmental changes and mediating adequate cellular responses. Inhibition of TOR complex 1 (TORC1) induces an immediate drop in the synthesis rate of ribosomes. It was previously suggested that TOR inactivation interferes with ribosome synthesis in many ways, but it was unclear whether and how these processes are coordinated.

To distinguish between primary and secondary effects on ribosome biogenesis in the yeast *Saccharomyces cerevisiae* and to determine the target mediating the fast response to TOR inactivation, Pol I transcription and rRNA synthesis were investigated shortly after TOR inhibition by rapamycin. This drug mimics nutrient starvation of cells by specifically inactivating the kinase activity of TORC1. The following conclusions could be drawn:

1) A rather long-term response constitutes the decrease in the level of Rrn3p leading to less initiation-competent Pol I-Rrn3p complex formation and thus reduced Pol I transcription. Rrn3p is characterized by a short half-life which is due to its constitutive ubiquitin-dependent degradation. Consequently, the level of Rrn3p is quickly down-regulated when the neo-synthesis of the protein is inhibited.

2) The fast down-regulation of mature rRNA synthesis correlates with serious pre-rRNA processing defects and subsequent RNA degradation, but not with the inhibition of Pol I transcription, since the association of Pol I with the rRNA gene locus is yet unaltered and the Pol I molecules engaged in transcription are still mobile.

3) The quick down-regulation of r-protein synthesis is sufficient to explain the severe pre-rRNA processing defects. The strong decrease in general translation, presumably along with the specifically reduced transcription rate of ribosomal protein genes, seems to cause the drastic repression of r-protein production.

SUMMARY

Since the level of Rrn3p appears to play a crucial role in Pol I transcription in yeast, this issue was investigated in more detail. Interestingly, already scarce amounts of Rrn3p are sufficient to promote Pol I transcription and cell growth, whereas strong overexpression of this factor results in growth defects. Elevated levels of Rrn3p lead to enhanced Pol I-Rrn3p complex formation, however, the question whether the growth defect is caused by the concomitantly observed increase in pre-rRNA-levels remains to be elucidated.

Finally, Pol5p, which was published to play an essential role in the synthesis of ribosomal RNA in yeast, co-purified with Rrn3p through several purification steps suggesting an interaction between the two proteins. However, further experiments provided only weak additional evidence for Pol5p as a genuine interaction partner of Rrn3p and failed to confirm the reported association of this protein with the rRNA gene locus. Therefore, further investigation is required to elucidate the role of Pol5p in ribosome biogenesis.

2 INTRODUCTION

2.1 Ribosome biogenesis and cell growth

The ability of cells to produce large amounts of proteins is indispensable for growth and proliferation, since proteins are required for almost every cellular process. Ribosomes are the molecular factories that carry out protein synthesis by translating the genetic code into the poly amino acid chains of proteins. Therefore, synthesis of ribosomes is one of the most important tasks of a growing cell (Rudra and Warner, 2004; Lempiäinen and Shore, 2009).

The eukaryotic ribosome is a ribonucleoprotein particle (RNP) consisting of two different subunits, the 40S small ribosomal subunit (SSU) and the 60S large ribosomal subunit (LSU) (Wilson and Nierhaus, 2003). These subunits in turn are comprised of four ribosomal RNA (rRNA) species and 79 ribosomal proteins (r-proteins/RPs). In yeast [mammals], the small ribosomal subunit is composed of the 18S rRNA and 33 r-proteins (rpS – ribosomal protein small subunit), whereas the 25S [28S], 5.8S and 5S rRNA along with 46 [49] r-proteins (rpL – ribosomal protein large subunit) form the large ribosomal subunit (Planta and Mager, 1998; Gerbasi et al., 2004; Moss et al., 2007).

Ribosome biogenesis requires the coordinated activity of all three nuclear RNA polymerases present in eukaryotic cells. A specialized RNA polymerase, RNA polymerase I (Pol I), is exclusively responsible for the transcription of a 35S [47S] precursor rRNA which is subsequently processed into the mature 18S, 5.8S and 25S [28S] rRNA species (see section 2.2). The transcription of the genes coding for ribosomal proteins is dependent on the activity of RNA polymerase II (Pol II), whereas the 5S rRNA gene is transcribed by RNA polymerase III (Pol III). In addition to the RNA and protein components of the ribosome itself, over 150 trans-acting ribosome biogenesis factors and about 100 small nucleolar RNAs (snoRNAs) participate in the complex maturation pathway of ribosomes (Figure 1) (Kressler et al., 1999; Venema and Tollervey, 1999; Fatica and Tollervey, 2002; Tschochner and Hurt, 2003).

Logarithmically growing cells of the budding yeast *Saccharomyces cerevisiae* harbor roughly 200000 ribosomes. Considering a generation time of 100 min, each cell has to produce 2000 ribosomes per min, thereby consuming a huge part of the cell´s energy. Strikingly, 60% of total transcription is devoted to ribosomal RNA and 50% of the RNA polymerase II-mediated transcription initiation events involve ribosomal protein genes (Warner, 1999). Hence, a precise and quick regulation of ribosome biogenesis in response to environmental changes is essential for the cell in order to avoid the waste of valuable resources. The target of rapamycin (TOR) (see section 2.3) and the protein kinase A (PKA) pathway are examples for signal transduction pathways that positively and/or negatively influence the ribosome biogenesis machinery and in turn cell growth with respect to the availability of nutrients, growth factors, carbon or nitrogen, respectively (Klein and Struhl, 1994; Thomas and Hall, 1997; Powers and Walter, 1999; Warner, 1999; Rudra and Warner, 2004).

INTRODUCTION

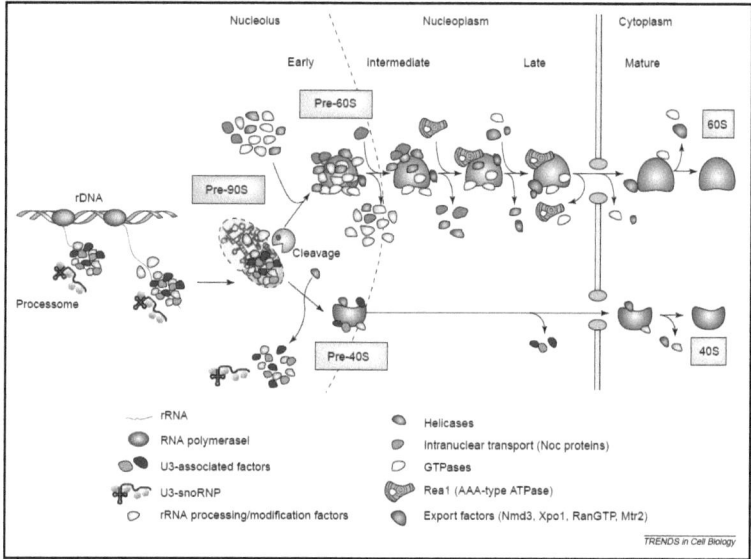

Figure 1. Overview of ribosome biogenesis in *Saccharomyces cerevisiae*.
The 35S pre-rRNA is assembled into the pre-90S particle which is separated into the pre-60S and the pre-40S particle upon cleavage of its rRNA component. During serveral maturation steps, these precursor particles are further processed into the mature 60S and 40S ribosomal subunits consisting finally of 4 rRNA species and 79 ribosomal proteins. More than 150 ribosome biogenesis factors and about 100 snoRNAs are transiently involved in this complex process. [from (Tschochner and Hurt, 2003)]

2.2 RNA polymerase I transcription in eukaryotes

2.2.1 Structure of ribosomal RNA genes

In eukaryotic cells, the synthesis of ribosomal RNAs is spatially restricted to a specialized compartment of the nucleus, the nucleolus, which is morphologically composed of the fibrillar centre, the dense fibrillar component and the granular component (Léger-Silvestre et al., 1999). Here, each cell of the yeast *Saccharomyces cerevisiae* holds about 100-140 copies of the genes coding for ribosomal RNA which are located in a tandemly repeated manner on chromosome XII (Schweizer et al., 1969; Petes, 1979). Mammalian cells contain 200-300 copies of ribosomal RNA genes per haploid genome which exist as direct repeats on the five acrocentric chromosomes (Henderson et al., 1972). Each of these sites has the potential to form a nucleolus and is hence referred to as a nucleolar organizer region (NOR). However, it was shown that in both lower and higher eukaryotes only about 50% of the chromosomal rDNA repeats are actively transcribed at any given time, whereas the other half of the rDNA genes is transcriptionally inactive (Conconi et al., 1989; Dammann et al., 1993; Moss, 2004).

In yeast, three of the four rRNAs, encoded by one 9.1 kb rDNA repeat, are transcribed by RNA polymerase I as a single polycistronic precursor, the 35S pre-rRNA. This transcript contains the mature rRNA sequences (18S, 5.8S and 25S), separated by the two internal transcribed spacers

INTRODUCTION

ITS1 and ITS2, and flanked by the two external transcribed spacers 5' ETS and 3' ETS. The remaining part of the rDNA unit is formed by the two non-transcribed spacers NTS1 and NTS2, separated by the 5S rRNA gene which is transcribed by RNA polymerase III in the opposite direction as the 35S rRNA gene (Figure 2). The 35S pre-rRNA is cleaved at the indicated sites (A_0 – E) in the course of subsequent processing and maturation steps to yield the mature rRNA species (see section 2.2.4) (Venema and Tollervey, 1999).

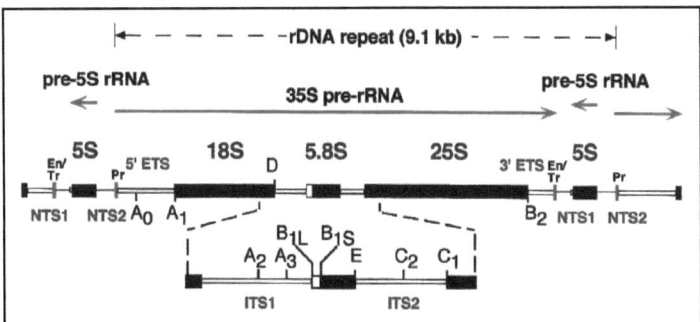

Figure 2. The basic organization of an rDNA repeat unit of *Saccharomyces cerevisiae*.
The yeast 9.1 kb rDNA unit consists of the 35S pre-rRNA operon and the two non-transcribed spacers NTS1 and NTS2, interrupted by the 5S rRNA gene. The 35S pre-rRNA contains the sequences for the mature 18S, 5.8S and 25S rRNAs, separated by the two internal transcribed spacers ITS1 and ITS2, and flanked by the two external transcribed spacers 5' ETS and 3' ETS. The locations of the known processing sites are indicated. (Pr: promoter, Tr: terminator, En: enhancer) [from (Kressler et al., 1999)]

The yeast *Saccharomyces cerevisiae* is the only known exception among eukaryotes, in which the 5S genes are linked to the rRNA genes and hence must necessarily be transcribed in the nucleolus. In mammalian cells, the 5S rRNA genes are transcribed from a different chromosomal location than the precursor transcripts containing the sequences of the mature 18S, 5.8S and 28S rRNAs and thus need to be imported into the nucleolus (Moss et al., 2007). However, the 47S pre-rRNA, transcribed from one 43 kb rDNA repeat, is subsequently processed to yield the mature rRNAs in a way very similar to that in yeast.

2.2.2 Subunit composition of RNA polymerase I

The yeast enzyme RNA polymerase I has a molecular weight of 590 kDa and is a multi-protein complex consisting of 10 different core-subunits and 4 different additional subunits (Carles et al., 1991; Carles and Riva, 1998; Kuhn et al., 2007). Their designation in the common Pol I nomenclature is composed of the letter A, B and/or C indicating the appearance of the subunit in RNA polymerase I, II and/or III, respectively, and of a number denoting the respective molecular weight in kDa as determined by SDS-PAGE (Table 1).
Five of the subunits, ABC27, ABC23, ABC14.5, ABC10β and ABC10α, are identical in all three nuclear polymerases (Carles et al., 1991). The two large subunits A190 and A135, comprising the active center of the enzyme, are unique to RNA polymerase I but contain regions homologous to

INTRODUCTION

the Pol II subunits Rpb1p and Rpb2p, respectively (Mémet et al., 1988). Furthermore, the two subunits AC40 and AC19 are common in Pol I and Pol III and share homologies with Rpb3p and Rpb11p, the corresponding subunits of RNA polymerase II (Lalo et al., 1993). The subunits A14 and A43 form a heterodimer which is distantly related to Rpb4p/Rpb7p in Pol II and Rpc17p/Rpc25p in Pol III (Peyroche et al., 2002; Geiger et al., 2008). A43 plays an important role in transcription initiation, since the basal transcription factor Rrn3p recruits RNA polymerase I to the promoter via this subunit (see section 2.2.3) (Milkereit and Tschochner, 1998; Peyroche et al., 2000). Subunit A12.2 is homologous to subunit Rpb9p in Pol II and Rpc11p in Pol III. Additionally, its C-terminal domain is related to the Pol II transcript cleavage factor TFIIS. This subunit indeed confers intrinsic RNA cleavage activity which is supposed both to enable rRNA proofreading and to play a major part in its role in efficient transcription termination (Prescott et al., 2004; Kuhn et al., 2007). No counterparts in other polymerases have been found for subunits A49 and A34.5. However, local homologies were detected between these two and the Pol II-associated factors TFIIF and TFIIE. It was shown that these subunits form a TFIIF-like heterodimer which provides a built-in elongation factor for RNA polymerase I (Kuhn et al., 2007; Geiger et al., 2010).

Yeast			Mammal/Human
Pol I subunit	gene locus	in Pol(s)	Pol I subunit orthologue
A190	RPA190	I	hRPA190 (A190, A194)
A135	RPA135	I	hRPA135 (A127)
A49	RPA49	I	hRPA49 (hPAF53)
A43	RPA43	I	hRPA43 (A43, TWIST neighbor)
AC40	RPA40	I, III	hRPA40 (AC40, hRPA5)
A34.5	RPA34.5	I	hRPA34.5 (hPAF49, CAST, ASE-1)
ABC27	RPB5	I, II, III	hRPB5
ABC23	RPB6	I, II, III	hRPB6
AC19	RPA19	I, III	hRPA19 (AC19)
ABC14.5	RPB8	I, II, III	hRPB8
A14	RPA14	I	-
A12.2	RPA12.2	I	hRPA12.2
ABC10β	RPB10	I, II, III	hRPB10
ABC10α	RPB12	I, II, III	hRPB12

Table 1. The 14 Pol I subunits of *Saccharomyces cerevisiae* and their mammalian/human orthologues.
See text for further explanation. [from (Panov et al., 2006b), modified]

Of all the 14 Pol I subunits in yeast, just four are not essential for cell growth, which are: A34.5, A49, A14 and A12.2. The respective deletion of the last three, however, leads to growth defects (Liljelund et al., 1992; Nogi et al., 1993; Smid et al., 1995).

INTRODUCTION

There are mammalian orthologues for all but yeast RNA Pol I subunit A14 (Table 1) (Panov et al., 2006b), indicating that the yeast enzyme constitutes a good model for studying eukaryotic RNA polymerase I.

2.2.3 RNA polymerase I transcription factors and their function

In all eukaryotes from yeast to mammals, the DNA elements directing Pol I transcription are very similar. The promoter region of each rDNA repeat unit consists of two *cis* elements: the upstream (control) element (UE/UCE) and the core element (CE/Core) (Figure 3 and Figure 4). The core element, mapped from about -40 to +8 relative to the transcription start site, is essential for both basal levels of transcription and accurate transcription initiation, whereas the upstream element, mapped from about -150 to -40 relative to the start site, is required for a high level of transcription, but is dispensable for transcription initiation *in vitro*. Interestingly, the maintenance of correct spacing between the two elements is critical (Musters et al., 1989; Kulkens et al., 1991; Choe et al., 1992; Paule, 1998). The terminator region in the 3' ETS of each rDNA repeat unit is characterized by the presence of a T-rich element and a further downstream binding site for a sequence-specific protein (Mason et al., 1997; Reeder and Lang, 1998).

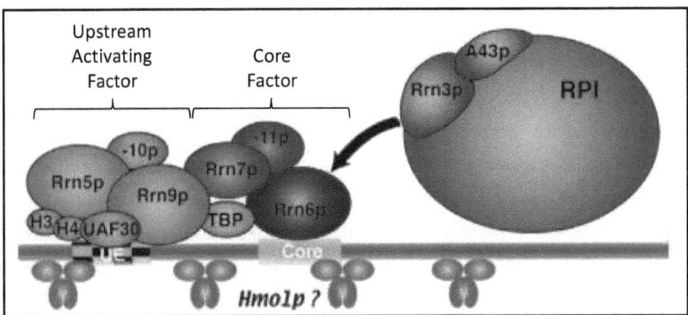

Figure 3. The RNA polymerase I initiation complex in *Saccharomyces cerevisiae*.
The upstream activating factor (UAF), consisting of 6 subunits, and the core factor (CF), consisting of 3 subunits, bind to the upstream element (UE) or the core element (Core), respectively, two characteristic features within the rDNA promoter region. The TATA-binding protein (TBP) forms a bridge between the two transcription factors thereby stabilizing this platform provided for the binding of RNA polymerase I. Pol I is recruited to the promoter via the essential transcription initiation factor Rrn3p interacting both with CF subunit Rrn6p and Pol I subunit A43. Binding of the HMG-box protein Hmo1p throughout the rRNA genes is required for efficient rDNA transcription. [from (Moss et al., 2007), modified]

In yeast, transcription initiation involves the coordinated interactions of at least four transcription factors with these promoter elements and RNA polymerase I: the upstream activating factor (UAF) (Keys et al., 1996; Keener et al., 1997), the core factor (CF) (Keys et al., 1994; Lalo et al., 1996; Lin et al., 1996), the TATA-binding protein (TBP) (Steffan et al., 1996, 1998) and Rrn3p (Yamamoto et al., 1996) (Figure 3).

The CF is a multi-subunit complex consisting of the three proteins Rrn6p, Rrn7p and Rrn11p which are all essential *in vivo*. It binds to the Core and is able to direct a basal level of Pol I

INTRODUCTION

transcription. The UAF, which interacts with the UE, constitutes a complex of six proteins including the four non-essential subunits Rrn5p, Rrn9p, Rrn10p and Uaf30p and the two histones H3 and H4. Contrary to the CF, the UAF is not absolutely required for specific initiation, but stimulates transcription by efficiently recruiting the CF to the promoter. The TBP, which interacts both with CF via Rrn6p and UAF via Rrn9p, appears to be necessary only for the UAF-dependent recruitment of the CF. In order to gain competence for initiation, RNA polymerase I forms a stable complex with the transcription initiation factor Rrn3p via its subunit A43. Rrn3p is crucial for recruiting the enzyme to the promoter by providing a bridge to the CF subunit Rrn6p. After transcription initiation, both TBP and CF dissociate from the promoter while UAF remains behind. Similarly, Rrn3p is released both from the promoter and the elongating form of RNA polymerase I. Pol I subunit A49 appears to play a crucial role in both the formation of the Pol I-Rrn3p complex and its subsequent dissociation (Milkereit and Tschochner, 1998; Peyroche et al., 2000; Aprikian et al., 2001; Bier et al., 2004; Beckouet et al., 2008).

Furthermore, binding of the factor Hmo1p throughout the complete rRNA gene locus is a prerequisite for efficient RNA Pol I transcription (Gadal et al., 2002). Since this high mobility group (HMG)-box protein is reported not only to associate with rRNA genes but also with many promoters of RP genes, it is speculated that it might function in coordinating the transcription of ribosomal RNA and ribosomal protein genes (Hall et al., 2006). Besides Hmo1p, other proteins are reported to play a role in efficient Pol I transcription elongation in yeast. For instance, the enzymatic activity of Fcp1p, a phosphatase originally described to be involved in Pol II transcription elongation, is likewise involved in the Pol I system (Fath et al., 2004). Similar results were obtained for Spt4p and Spt5p. This heterodimer influences both Pol II and Pol I transcription elongation. Deletion of the non-essential gene for Spt4p leads also to clear defects in pre-rRNA processing, indicating that transcription and processing are intimately linked (Schneider et al., 2006). RNA polymerase-associated factor 1 complex (Paf1C), a complex composed of five subunits, was recently shown to promote Pol I transcription through the rDNA by increasing the net rate of elongation (Zhang et al., 2009, 2010). Additionally, Net1p, which forms the regulator of nucleolar silencing and telophase exit (RENT) complex along with at least Cdc14p and Sir2p, was described to mediate high rates of Pol I transcription besides its roles in controlling mitotic exit and diverse other nucleolar processes (Shou et al., 2001). Another example for factors involved in the regulation of Pol I transcription elongation is Ctk1p, the kinase subunit of a complex described to participate in the regulation of mRNA synthesis by Pol II (Bouchoux et al., 2004).

Results from *in vitro* studies suggest that accurate Pol I transcription termination in yeast depends on the binding of the essential protein Reb1p to its target sequence within the terminator region which forces the polymerase to pause. This, along with the weak base pairing between template DNA and RNA transcript within the T-rich element, leads to the release of both the polymerase and the transcript from the DNA template (Lang et al., 1994; Lang and Reeder, 1995). It was further shown that Pol I depends on an additional factor to release

INTRODUCTION

terminated transcripts from the template (Tschochne and Milkereit, 1997). In an alternative model of transcription termination in yeast, co-transcriptional cleavage of the pre-rRNA by the endonuclease Rnt1p is proposed to generate a loading site for the exonuclease Rat1p which degrades the nascent transcript from the 5' end and finally torpedoes the polymerase (El Hage et al., 2008; Kawauchi et al., 2008; Braglia et al., 2010).

Although distinct functional similarities between the yeast and the mammalian Pol I transcription system are obvious, there are nevertheless certain differences regarding the factors involved.

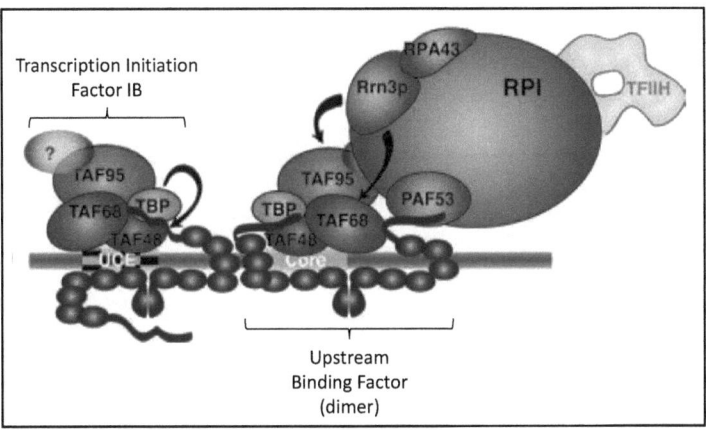

Figure 4. The RNA polymerase I initiation complex in mammals.
The upstream binding factor (UBF) binds probably as a dimer both to the upstream control element (UCE) and the core element (Core) and creates a situation propitious for selectivity factor 1 (SL1)/transcription initiation factor IB (TIF-IB), consisting of at least 4 subunits, to bind and to form a stable pre-initiation complex. Pol I recruitment is accomplished by the human RRN3 (hRRN3)/transcription initiation factor IA (TIF-IA) which interacts both with Pol I subunit A43 and SL1/TIF-IB subunits TAF$_I$63/68 or TAF$_I$110/95, respectively. TFIIH is additionally required for productive rDNA transcription. [from (Moss et al., 2007), modified]

In mammals, the human selectivity factor 1 (SL1) (Learned et al., 1985) or the mouse transcription initiation factor IB (TIF-IB) (Clos et al., 1986), respectively, in combination with the upstream binding factor (UBF) (Jantzen et al., 1990) is required to promote efficient transcription initiation by providing a platform to which RNA polymerase I is recruited via the human RRN3 (hRRN3) (Moorefield et al., 2000) or the mouse transcription initiation factor IA (TIF-IA) (Bodem et al., 2000), respectively (Figure 4).

SL1 is composed of TBP and at least three TBP-associated factors (TAFs), including TAF$_I$48, TAF$_I$63 and TAF$_I$110. TIF-IB exhibits the same composition except for the last two TAFs being named TAF$_I$68 and TAF$_I$95. These essential TAFs are apparently the mammalian orthologues to the three CF subunits in yeast. *In vitro*, SL1/TIF-IB is sufficient to provoke basal levels of transcription by Pol I. Activated transcription, however, also requires the non-specific DNA-binding protein UBF which in part resembles yeast Hmo1p and UAF. UBF binds as a dimer to the UCE and the Core via its HMG-boxes. SL1/TIF-IB subunits TAF$_I$48 and TBP interact with the highly acidic C-terminus of

INTRODUCTION

UBF, thereby recruiting SL1 to the promoter. Similarly to the situation in yeast, the essential factor hRRN3/TIF-IA brings RNA polymerase I to the promoter by forming a bridge between the Pol I subunit A43 and the two SL1/TIF-IB subunits $TAF_I63/68$ or $TAF_I110/95$, respectively. After initiation, Pol I escapes the promoter and converts into the elongating form which coincides with the loss and inactivation of hRRN3/TIF-IA (Grummt, 2003; Moss, 2004; Russell and Zomerdijk, 2005; Moss et al., 2007).

Besides its role in transcription initiation, UBF is also shown to play a role in correct promoter escape and in efficient transcription elongation (Stefanovsky et al., 2006; Panov et al., 2006a). Similarly, the facilitates chromatin transcription (FACT) complex stimulates elongation by facilitating Pol I transcription through nucleosomal templates (Birch et al., 2009). Furthermore, two Pol II transcription factors are reported to be involved in the Pol I system. On the one hand, there is no productive transcription in the absence of TFIIH, which implies a post-initiation role for this multi-subunit complex. On the other hand, it was suggested that the RNA cleavage activity mediated by TFIIS is required for both Pol II and Pol I to overcome transcriptional impediments during RNA chain elongation (Schnapp et al., 1996; Iben et al., 2002). It should be noted, however, that such a role for TFIIS in the yeast Pol I system was not detected in a different study (Tschochner, 1996).

Accurate transcription termination in mammals depends also on auxiliary factors. Here, binding of the transcription termination factor I (TTF-I) to its target site in the 3' ETS induces DNA bending and pausing of Pol I. TTF-I cooperates with the polymerase and transcript release factor (PTRF) in conjunction with the T-rich element to mediate transcription termination and dissociation of both the elongating Pol I and the transcript from the template (Jansa and Grummt, 1999; Russell and Zomerdijk, 2005). To date, no alternative model for Pol I transcription termination has been proposed in mammals resembling the yeast torpedo-mechanism.

2.2.4 Pre-rRNA processing and rRNA maturation

As mentioned above, the 18S, 5.8S and 25S [28S] ribosomal RNA species in yeast [mammals] are transcribed by RNA polymerase I as a single polycistronic precursor, the 35S [47S] pre-rRNA, which is subsequently matured in a complex series of co- and post-transcriptional processing steps to yield the mature RNAs.

In yeast, a subset of ribosomal and non-ribosomal proteins along with diverse small nucleolar ribonucleoprotein particles (snoRNPs) assemble to the precursor rRNA in the course of transcription to establish the initial 90S pre-ribosomal particle. Co-transcriptional cleavage at site B_0 within the 3' ETS releases the pre-90S particle containing the 35S pre-rRNA which is target of extensive processing events and modification processes comprising mainly 2' O-methylations and pseudouridylations.

Subsequent processing at sites A_0 and A_1 in the 5' ETS of the 35S pre-rRNA yields the 32S pre-rRNA which is endonucleolytically cleaved at site A_2 in the ITS1 to give birth to the pre-40S and pre-60S particle containing either the 20S or the $27SA_2$ pre-rRNA species, respectively. The pre-

INTRODUCTION

40S particle is exported to the cytoplasm where it is converted into the mature small ribosomal subunit by cleavage of the 20S pre-rRNA at site D producing the 18S rRNA (Figure 5).

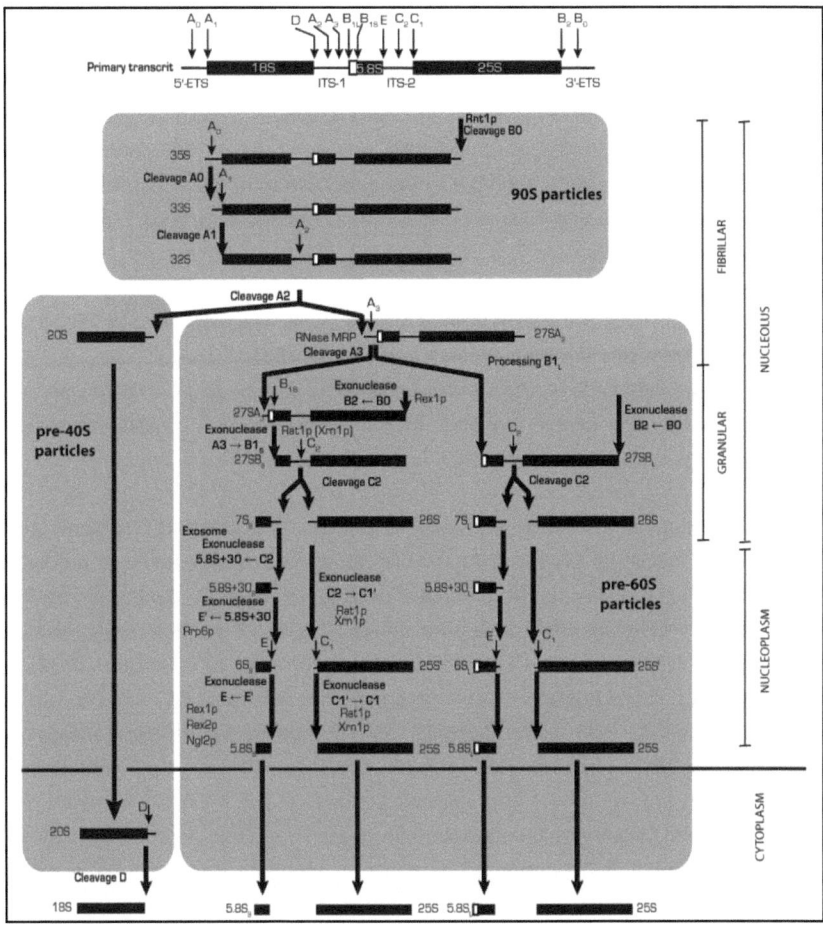

Figure 5. Overview of pre-rRNA processing pathways in *Saccharomyces cerevisiae*.
The upper panel shows a schematic drawing of the 35S pre-rRNA transcript with the locations of the respective processing sites. The central panal depicts the successive processing steps from the 35S to the 32S pre-rRNA within the pre-90S particle. An endonucleolytic cleavage event separates the processing pathways of the pre-40S and the pre-60S particle, both of which are illustrated in the two lower panels. Subsequent conversion of the 20S pre-rRNA and the 27SA$_2$ pre-rRNA into the mature rRNA species is shown. The intermediate rRNA species and the implications of diverse exo- and endonucleolytic cleavage activities are depicted, as are the cell compartments where the respective processing steps occur. [from (Henras et al., 2008)]

Processing of the 27SA$_2$ pre-rRNA occurs by two alternative pathways. The major pathway involves cleavage at site A$_3$ in the ITS1 producing the 27SA$_3$ pre-rRNA. This is very rapidly followed by exonucleolytic digestion to site B$_{1S}$ forming the mature 5' end of the short form of the 5.8S rRNA (5.8S$_S$) within the 27SB$_S$ pre-rRNA. The second, minor pathway is characterized by a processing step at site B_{1L} which creates the $27SB_L$ pre-rRNA harboring the mature 5' end of the long form of the 5.8S rRNA ($5.8S_L$). Exonucleolytic digestion at site B_2 generates the 3' end of the mature 25S rRNA. The subsequent processing steps of the $27SB_S$ and $27SB_L$ pre-rRNA species appear to be identical. Both are cleaved at site C_2 within the ITS2. The released upstream fragments, $7S_S$ and $7S_L$, are processed at their 3' ends by diverse exoribonucleases, thereby forming the $6S_S$ and $6S_L$ pre-rRNA species which are further processed to the mature $5.8S_S$ and $5.8S_L$ rRNAs. Finally, the 5' end of the mature 25S rRNA is obtained by exonucleolytic digestion to site C_1. Contrary to the pre-40S particle, all rRNA species of the pre-60S particle are matured completely before the particle is exported to the cytoplasm to function as the large ribosomal subunit (Figure 5) (Venema and Tollervey, 1999; Tschochner and Hurt, 2003; Fromont-Racine et al., 2003; Henras et al., 2008).

It is noteworthy that eukaryotic cells contain at least two different types of ribosomes, possessing either the long or the short form of the 5.8S rRNA, which may in principle be capable of translating different sets of mRNAs (Schmitt and Clayton, 1993).

Although there are differences in the pre-rRNA processing and modification pathways between yeast and mammals, the overall sequence of maturation events in eukaryotes seems very related, since trans-acting factors involved in ribosome biogenesis are highly conserved (Henras et al., 2008).

Importantly, continuous availability of ribosomal proteins in at least stoichiometric amounts with the rRNA is crucial for proper maturation of ribosomal subunits. Reduced production of individual ribosomal proteins due to conditional depletion or r-protein gene haploinsufficiency rapidly leads to severe pre-rRNA processing defects (Lucioli et al., 1988; Song et al., 1996; Deutschbauer et al., 2005; Ferreira-Cerca et al., 2005; Robledo et al., 2008; Pöll et al., 2009).

It was suggested that the primary 35S pre-rRNA transcript most probably starts to fold and to interact with snoRNAs and both ribosomal and non-ribosomal proteins already during transcription. Several years ago, the opinion prevailed that pre-rRNA processing and modification does not commence until cleavage at site B_0 in the 3' ETS is completed (Venema and Tollervey, 1999). However, although RNA polymerase I transcription proceeds in some cases unabated until the 3' ETS is synthesized, the nascent transcript could also be modified and cleaved co-transcriptionally in the ITS1, thereby immediately releasing a pre-40S particle without prior pre-90S particle formation (Osheim et al., 2004; Kos and Tollervey, 2010). This observation along with further recent findings suggest that rRNA gene transcription and the downstream pre-rRNA processing events are intimately linked (Granneman and Baserga, 2005). Indeed, accurate transcription elongation by RNA Pol I is a prerequisite for efficient pre-rRNA processing and pre-ribosome assembly (Schneider et al., 2007). Similarly, two independent studies revealed a subset of early assembling non-ribosomal proteins to be implicated not only in accurate pre-rRNA processing but also in efficient rDNA transcription (Gallagher et al., 2004; Prieto and McStay, 2007). Further evidence for a connection between these processes derived from the

INTRODUCTION

observation that depletion of the Pol II elongation factor Spt4p, which is similarly involved in the Pol I system, results in a pre-rRNA processing defect (Schneider et al., 2006).

In the next section, two proteins are introduced which are potential candidates for coordinating rDNA transcription and pre-rRNA processing.

2.2.5 Pol5p and Mybbp1a – potential regulators of ribosomal RNA synthesis

Pol5p, a constitutively expressed 116 kDa protein, was originally characterized to be the fifth essential DNA polymerase in *Saccharomyces cerevisiae* and is therefore named DNA polymerase φ (Shimizu et al., 2002). However, although this factor both exhibits significant DNA polymerization activity *in vitro* and contains each of the six characteristic Pol domains that are present in all B-type polymerases, its role as a *bona fide* DNA polymerase is still discussed controversially (Yang et al., 2003). Mutational analysis showed that Pol5p plays an essential role in a cellular function other than chromosomal DNA replication. Since Pol5p co-localizes exclusively with the nucleolar marker protein Nop1p, a role in regulating the synthesis of ribosomal RNA seemed to be possible. Indeed, temperature-sensitive *POL5* mutant strains rapidly ceased growth, displayed severe inhibition of rRNA synthesis and increased the number of rDNA repeat units on chromosome XII at the restrictive temperature. Additionally, cultivation of these mutant cells at the restrictive temperature led to an altered cellular distribution pattern of Pol5p. Instead of nucleolar localization, the factor was now detected in punctate foci in the cytoplasm (Shimizu et al., 2002). Furthermore, association of Pol5p with the rRNA gene locus was shown by chromatin immunoprecipitation experiments pointing to a function in rDNA transcription. Various studies report specific crosslinking of Pol5p to the promoter region, the 25S rRNA-coding region and the rDNA enhancer region (Shimizu et al., 2002; Nadeem et al., 2006; Wery et al., 2009). The latter is located in the 3' ETS and overlaps with the Reb1p-binding site which is important for transcription termination. The enhancer element was shown to exhibit a stimulatory effect on rRNA synthesis by Pol I in *in vitro* and *in vivo* experiments from Pol I reporter templates (Elion and Warner, 1984, 1986), but is dispensable for rDNA transcription in the chromosomal context *in vivo* (Wai et al., 2001). Moreover, Pol5p was recently identified within a complex that early assembles to the 35S pre-rRNA suggesting a function in pre-rRNA maturation (Krogan et al., 2004).

In summary, these data strongly indicate a role for Pol5p in rRNA synthesis, though it is still unknown whether this protein participates predominantly in rDNA transcription, in pre-rRNA processing or in both, thereby potentially coordinating these essential events concerning ribosome production.

In mammals, Myb-binding protein 1a (Mybbp1a) is considered to be the corresponding homologue of Pol5p. The designation of this ubiquitously expressed and predominantly nucleolar protein initially derived from its ability to interact with the proto-oncogene product c-Myb, a transcription factor that is critical for cell proliferation and differentiation (Favier and Gonda, 1994; Tavner et al., 1998; Keough et al., 2003). To date, the function of Mybbp1a has been mainly described in the context of RNA polymerase II-dependent transcription. Nevertheless, very recent findings support the speculation for Mybbp1a to be involved in the Pol I transcription system as well. Strong evidence exist that Mybbp1a interacts with Pol I subunit hPAF53/A49. Additional experiments in this study revealed that this protein serves both as a negative regulator of rRNA gene transcription and as a functional subunit of the ribosome biogenesis machinery, thereby influencing pre-rRNA processing (Hochstatter et al., submitted). This dual role in the rDNA metabolism points to Mybbp1a to be a coordinator of rDNA transcription and pre-rRNA processing.

Taken together, lots of factors are evidently involved in Pol I transcription and/or pre-rRNA processing. However, it is not clear yet, which of these factors are influenced by the essential TOR pathway in order to precisely regulate the complex process of ribosome biogenesis in response to environmental changes.

2.3 TOR – a central component of the eukaryotic growth regulatory network

2.3.1 General description of the target of rapamycin (TOR)

In the 1970s, a potent antifungal metabolite was discovered which was produced by the bacterial strain *Streptomyces hygroscopicus* isolated from a soil sample from Easter Island, locally known as Rapa Nui (Vézina et al., 1975; Sehgal et al., 1975). This macrocyclic lactone, which was named rapamycin after its place of discovery, showed immunosuppressive properties and inhibited proliferation of mammalian cells. Rapamycin was further investigated to elucidate its mode of action. During these studies, the target of rapamycin (TOR) was originally identified by the mutations *tor1-1* and *tor2-1* in *Saccharomyces cerevisiae* that confer resistance to the growth inhibitory properties of rapamycin. Concomitantly, availability of an intracellular co-factor was described to be crucial for rapamycin toxicity (Heitman et al., 1991). Prior to binding and thus inhibiting TOR, rapamycin has to form a complex with the peptidyl-prolyl *cis/trans* isomerase FKBP12.

To date, every eukaryote genome examined contains a *TOR* gene. Most higher eukaryotes possess a single *TOR* gene, whereas *Saccharomyces cerevisiae* holds two of these genes. All the proteins encoded by such genes belong to a group of kinases known as the phosphatidylinositol kinase-related kinase (PIKK) family, the members of which share distinct domain features (Figure 6). A ser/thr-kinase domain confers the respective enzymatic activity. The matter of the FKBP12-rapamycin-binding (FRB) domain is already explained by its designation. FAT and FATC domains are supposed to interact within the protein, whereas the tandem HEAT repeats may provide interfaces for protein-protein interactions (Wullschleger et al., 2006).

The target of rapamycin constitutes a conserved cellular regulator of cell growth and metabolism which controls these complex processes in response to environmental changes via its kinase activity. Studies in yeast demonstrated that TOR performs essentially two distinct functions: it controls both when a cell grows and where a cell grows (Loewith and Hall, 2004). Two distinct TOR complexes are involved to accomplish this challenging task in the cell: TOR complex 1 (TORC1) and TOR complex 2 (TORC2) (Figure 6). The structure and function of these complexes are conserved from yeast to mammals. Interestingly, just TORC1 is sensitive to rapamycin treatment, whereas TORC2 is characterized by rapamycin resistance.

In yeast, TOR complex 1 consists of four proteins which are: Kog1p, Tco89p, Lst8p and either Tor1p or Tor2p. Interactions between the factors Avo1p, Avo2p, Avo3p, Lst8p, Bit61p and Tor2p constitute TOR complex 2 (Figure 6) (Loewith et al., 2002). Both TORC1 and TORC2 as well as the mammalian equivalents are supposed to be oligomeric supercomplexes, most likely dimers, based on the interaction between the respective Tor proteins (Figure 6) (Wullschleger et al., 2005).

When nutrients are available, signaling by the active TOR complex 1 regulates temporal aspects of cell growth by providing a robust rate of ribosome biogenesis, translation initation and nutrient import. Treatment of the cells with rapamycin, however, amino acid starvation or exposure of the cells to any other form of stress, inactivates TORC1 leading to a rapid down-regulation of general protein synthesis and concomitantly to an activation of both macroautophagy and stress-responsive transcription factors. Thus, the cells abruptly arrest growth and enter a G0-like state. In contrast, TORC2 controls spatial aspects of cell growth by organizing the polarization of the actin cytoskeleton through a rapamycin-insensitive signaling branch (Figure 6) (Wullschleger et al., 2006; De Virgilio and Loewith, 2006a, 2006b).

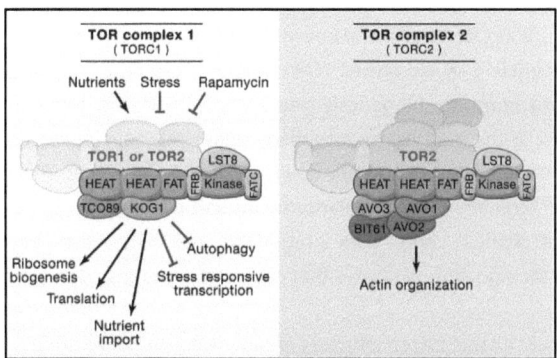

Figure 6. TOR complex 1 (TORC1) and TOR complex 2 (TORC2) of *Saccharomyces cerevisiae*.
The protein composition (Kog1p, Tco89p, Lst8p, Avo1-3p, Bit61p and Tor1p or Tor2p) of TOR complex 1 and TOR complex 2 is depicted, as is the domain organization (HEAT, FAT, FRB, Kinase, FATC) of Tor1p or Tor2p, respectively. Both complexes are oligomers, likely dimers. Rapamycin-sensitive TORC1 regulates growth in dependency of environmental conditions, whereas rapamycin-resistant TORC2 is involved in the organization of the actin cytoskeleton. Stimuli that activate TORC1 and TORC1 outputs that promote growth as well as the scope of TORC2 are illustrated with black arrows. Inputs that inhibit TORC1 signaling and processes that are negatively regulated by TORC1 are depicted with red bars. [from (Wullschleger et al., 2006)]

In mammals, the corresponding equivalents, mTORC1 and mTORC2, exhibit virtually the same characteristics and functions as their yeast homologues. Rapamycin-sensitive mTORC1 appears likewise to regulate the temporal aspects of cell growth, whereas mTORC2 controls the spatial aspects of cell growth in a rapamycin-resistant manner (Wullschleger et al., 2006).

In metazoans, TOR primarily controls growth during development, but in the adult, where cell growth plays a rather minor role, TOR controls aging and other aspects of nutrient-related physiology. Interestingly, partial inhibition of TOR function in worms and flies, but also in yeast, results in a significant increase in the life span of these organisms (Martin and Hall, 2005; Kaeberlein et al., 2005).

2.3.2 Upstream and downstream of the TOR signaling network

TOR integrates various signals to regulate cell growth. Four major inputs are considered to be involved in modulating TOR signaling in eukaryotic cells, which are: hormones and growth factors, nutrients, energy and stress (Wullschleger et al., 2006).

In mammals, several signaling pathways are shown to be implicated in both increasing and decreasing the activity of mammalian TOR complex 1. In most of these cases, the heterodimer TSC1-TSC2 acts as a central integration unit and conveys the essential information to mTORC1. Mammalian TOR complex 2 might also act downstream of TSC1-TSC2, but further investigation is required to elucidate this issue. To date, no upstream regulators of mTORC2 are reported (Wullschleger et al., 2006; De Virgilio and Loewith, 2006b).

In yeast, however, distinct upstream regulators of both TOR complex 1 and TOR complex 2 remain elusive to this day. It is noteworthy that no orthologues of TSC1 and TSC2 are identified yet in yeast indicating that upstream signals may be sensed differently in this organism (Wullschleger et al., 2006; De Virgilio and Loewith, 2006b).

Generally, active signaling by eukaryotic TOR complex 1 promotes overall protein synthesis via stimulation of both ribosome biogenesis (see section 2.3.3) and translation as well as other anabolic processes. In contrast, macroautophagy, other catabolic processes and the activity of stress-responsive transcription factors are down-regulated. Eukaryotic TOR complex 2 primarily plays a role in the regulation of cell polarity by organizing the actin cytoskeleton. Although a large number of TORC1- and TORC2-regulated readouts have been elucidated, the understanding of the signaling pathways that couple TOR to these downstream targets remains limited.

In mammals, the best characterized effectors of rapamycin-sensitive mTORC1 are the translation regulators ribosomal protein S6 kinase 1 (S6K1) and eIF4E-binding protein 1 (4E-BP1) (Hay and Sonenberg, 2004).

Full activation of S6K1, which is a member of the AGC kinase family, is achieved by the phosphorylation of two sites within distinct protein domains. Upon phosphorylation by both pyruvate dehydrogenase kinase 1 (PDK1) and mTORC1, S6K1 is activated and in turn phosphorylates ribosomal protein S6. Apparently, the posttranslational modification of this factor leads to an increase in the translation of a subset of mRNAs which contain a specific tract of oligopyrimidine at the 5' end (5' TOP). Since this kind of mRNAs encodes predominantly components of the translation apparatus such as ribosomal proteins and elongation factors, active S6K1 is thought to up-regulate general translation capacity. However, recent work has provided new evidence that neither S6K1/S6K2 activity nor phosphorylation of ribosomal protein S6 is required for mTORC1 to alter the translation efficiency of 5' TOP mRNAs. Furthermore, S6K1 appears to promote translation elongation by inhibiting the phosphorylation of the eukaryotic elongation factor 2 (eEF2). In a similar cascade, mTORC1 controls cap-dependent translation initiation via the translational inhibitor 4E-BP1. Phosphorylation of 4E-BP1 by mTORC1 leads to its dissociation from the eukaryotic initiation factor 4E (eIF4E) which is then free to associate with its target eIF4G to stimulate translation initiation.

Besides translation, mTORC1 signaling is further involved in macroautophagy, a cellular process by which cytoplasmic contents, including organelles, are degraded and thereby recycled in the vacuole in order to ensure the survival of the cell when nutrients are scarce. Activity of this catabolic pathway is largely dependent on mTORC1 signaling. Inactivation of mTORC1 by starvation conditions leads to enhanced macroautophagy, whereas under favorable growth conditions, this process is largely impaired.

Moreover, mTORC1 regulates many aspects of cellular metabolism including amino acid biosynthesis, glucose homeostasis and others. In particular, mTORC1 and S6K1 appear to play an important role in the fat metabolism. Loss of mTORC1 activity correlates with a significant decrease in fat accumulation highlighting the economical nature of this pathway.

The involvement of mTORC1 signaling in the transcriptional regulation of many genes, in the trafficking and activation of numerous nutrient transporters and in mRNA stability was suggested, but is also much less well characterized (Wullschleger et al., 2006; De Virgilio and Loewith, 2006a, 2006b; Soulard et al., 2009).

Rapamycin-resistant mTORC2 controls the cell cycle-dependent polarization of the actin cytoskeleton. Although the exact mechanism by which the mammalian TOR complex 2 signals to the actin cytoskeleton is still unknown, it appears that PKB/Akt, RHO family GTPases and protein kinase C (PKC) play essential roles in this process (Wullschleger et al., 2006; De Virgilio and Loewith, 2006a, 2006b; Soulard et al., 2009).

In yeast, the best studied direct downstream targets of TORC1 are the essential protein Tap42p and the non-essential AGC family kinase Sch9p (Di Como and Arndt, 1996; Jiang and Broach, 1999; Urban et al., 2007; Huber et al., 2009). Whereas Sch9p is predominantly involved in the regulation and coordination of ribosome biogenesis (see section 2.3.3), Tap42p-dependent TORC1 signaling controls primarily the localization and activity of various transcription factors. Additionally, both Sch9p and Tap42p participate in the regulation of translation.

In rapidly growing cells, Tap42p is phosphorylated by TORC1 and associates tightly with the oligomeric complexes of both type 2A (PP2A) and type 2A-related (PP2Ar) protein phosphatases. Impaired TOR signaling, however, results in the dephosphorylation of Tap42p and thus reduced interaction with both PP2A and PP2Ar. By controlling this switch, TORC1 modulates several transcription factors such as Gln3p. Dephosphorylation of this protein by active PP2A and PP2Ar

INTRODUCTION

in nitrogen starvation conditions causes its dissociation from the cytoplasmic repressor protein Ure2p and its translocation in the nucleus. Hereupon, the expression of nitrogen-catabolite repression (NCR)-sensitive genes enables the cell to use poor nitrogen sources. The heterodimeric transcription factor Rtg1p-Rtg3p is likewise tethered to the cytoplasm by TORC1 in growing cells. The retrograde response pathway signals mitochondrial dysfunction to TORC1 whose inactivation leads to the dephosphorylation of both Rtg1p-Rtg3p-heterodimer and its cytoplasmic anchor protein Mks1p. The activated transcription factor is now capable of inducing the expression of the corresponding genes in the nucleus in order to overcome this dysfunction. Msn2p and Msn4p, two more transcription factors, regulate stress-responsive element (STRE)-dependent transcription in response to a wide range of stresses. Active TORC1 maintains the phosphorylated and thus phosphatase-inhibiting status of Tap42p which in turn ensures the phosphorylation of Msn2p and Msn4p and their accumulation in the cytoplasm.

Besides regulating transcription factors, both TORC1 and Tap42-dependent PP2A/PP2Ar play a role in the degradation of amino acid transporters and in cell wall integrity by controlling the activity of either Npr1p or kinase Mpk1p, respectively.

Moreover, TORC1 also modulates the cellular protein synthesis rate both via Tap42p and Sch9p. In growing cells, the phosphorylated form of kinase Gcn2p is supported by the activity of kinase Sch9p along with the PP2A/PP2Ar-inhibitory function of phosphorylated Tap42p. Keeping Gcn2p phoshoprylated prevents phosphorylation and thus inactivation of the eIF2 subunit α ensuring a robust rate of translation initiation. Like its putative mammalian orthologue S6K1, Sch9p phosphorylates the respective yeast ribosomal protein S6, thereby positively regulating translation in a TORC1-dependent manner. Another positive effect on translation represents the diminished degradation rate of eIF4G which seems to be accomplished by TORC1 without the involvement of either Tap42p or Sch9p.

Interestingly, additional Tap42p- and Sch9p-independent TORC1 signaling pathways exist in the cell. Active TORC1 signaling promotes hyperphosphorylation of the protein Atg13p in growing cells which prevents its interaction with Atg1p and thus the initiation of macroautophagy. A further target of the TOR signaling pathway is the protein Ime1p. The function of this transcription factor is crucial in diploid cells to initiate meiosis and sporulation in response to unfavorable growth conditions by inducing a transcriptional cascade of sporulation-specific genes in the nucleus (Wullschleger et al., 2006; De Virgilio and Loewith, 2006a, 2006b; Soulard et al., 2009; Urban et al., 2007; Huber et al., 2009).

In contrast to TOR complex 1, the readouts downstream of TOR complex 2 in yeast are less well characterized. Currently the hypothesis prevails that TORC2-dependent activation of the GTPase Rho1p activates the AGC family kinase Pkc1p which in turn signals to the actin cytoskeleton via a subsequent phosphorylation cascade including kinases Bck1p, Mkk1p/Mkk2p and Mpk1p (Wullschleger et al., 2006; De Virgilio and Loewith, 2006a, 2006b; Soulard et al., 2009).

2.3.3 TOR signaling in the context of growth-dependent regulation of ribosome biogenesis

As already mentioned, ribosome biogenesis is an energetically very costly process in the cell that has to be tightly regulated in response to nutrient and energy conditions. RNA polymerase I transcription plays an essential role in this process, since it synthesizes most of the ribosomal RNA. Logarithmically growing yeast cells display high rRNA synthesis rates, whereas stationary yeast cells almost completely lack the production of ribosomal RNAs (Ju and Warner, 1994). Similarly, nutrient deprivation or rapamycin treatment of eukaryotic cells, which both leads to the inhibition of TOR signaling, results in a rapid decrease in Pol I transcription rates (Grummt et al., 1976; Zaragoza et al., 1998; Powers and Walter, 1999). Therefore, the activity of RNA polymerase I and thus ribosome biogenesis is apparently strictly regulated in a TOR-dependent manner.

RNA polymerase I is present in two distinct populations in both lower and higher eukaryotic cells (Bateman and Paule, 1986; Tower and Sollner-Webb, 1987; Milkereit et al., 1997; Miller et al., 2001). Both populations are capable of unspecifically synthesizing RNA *in vitro*, but only one is able to initiate at the rDNA promoter in cell-free transcription systems. As mentioned before, the initiation-competent population of Pol I is characterized by the fact of existing in a complex with the transcription factor Rrn3p or hRRN3/TIF-IA, respectively (Milkereit and Tschochner, 1998; Miller et al., 2001; Yuan et al., 2002). Interestingly, Pol I-Rrn3p complexes are exclusively detectable in growing cells, whereas in stationary cells and in cells starved for amino acids or treated with the protein synthesis inhibitor cycloheximide, this specialized Pol I fraction is largely absent (Buttgereit et al., 1985; Bateman and Paule, 1986; Tower and Sollner-Webb, 1987; Milkereit and Tschochner, 1998).

Consistently, in yeast cells following rapamycin-induced TOR inactivation, the amount of Pol I-Rrn3p complexes is decreased as is the association of Pol I with both the promoter and the transcribed region of the rDNA locus, nicely resembling the situation in stationary phase (Claypool et al., 2004). These observations suggest that in yeast the rate of Pol I transcription is strongly dependent on the formation of Pol I-Rrn3p complexes. *In vitro* experiments using transcriptional-inactive extracts of yeast or mammalian cells, respectively, showed that Rrn3p is only capable of restoring promoter-dependent Pol I transcription when it is bound to Pol I, whereas both recombinant and purified active TIF-IA by itself is sufficient to obtain the same result (Buttgereit et al., 1985; Milkereit and Tschochner, 1998; Yuan et al., 2002). Thus, in mammalian cells, the rate of Pol I transcription appears to be rather dependent on the activity of hRRN3/TIF-IA.

Since in yeast Rrn3p as well as the five Pol I subunits A190, A43, A34.5, ABC23 and AC19 are described to be phosphorylated *in vivo* (Bell et al., 1976, 1977; Buhler et al., 1976; Bréant et al., 1983; Fath et al., 2001), TOR signaling was speculated to influence the formation of Pol I-Rrn3p complexes via phosphorylation-dephosphorylation cascades in a growth-dependent manner. Indeed, *in vitro* experiments suggest that Pol I needs to be phosphorylated for binding to Rrn3p, whereas the latter is able to bind to Pol I in its unphosphorylated form. In addition, the free population of Rrn3p, accounting for roughly 75% of the total protein, is predominantly phosphorylated *in vivo* (Fath et al., 2001; Bier et al., 2004). Similarly, the 2% of total Pol I being associated with Rrn3p display a different phosphorylation pattern than the excess of unbound Pol I (Milkereit and Tschochner, 1998; Fath et al., 2001). However, the kinase and phosphatase activities responsible for these posttranslational modifications remain elusive to this day. Interestingly, recent findings in yeast report that kinase Tor1p is dynamically distributed in the cytoplasm and in the nucleus. The nuclear localization is shown to be critical for 35S pre-rRNA synthesis, which is consistent with the fact that Tor1p is associated with the rDNA promoter region in a nutrient-dependent and rapamycin-sensitive manner (Li et al., 2006). It is thus possible that Rrn3p but also other factors of the Pol I transcription machinery are direct targets of TORC1. Notably, nuclear-cytoplasmic shuttling was also reported for the mammalian TOR complex 1 (Kim and Chen, 2000).

In mammals, predominantly the phosphorylation status of hRRN3/TIF-IA appears to determine the ability of forming a complex with Pol I. *In vitro* transcription assays demonstrate that only phosphorylated hRRN3/TIF-IA is capable of binding to Pol I in order to promote transcription initiation which leads concomitantly to its dephosphorylation and thus its inability to reinitiate both complex formation and transcription (Cavanaugh et al., 2002; Hirschler-Laszkiewicz et al., 2003). Subsequent studies successfully identified several regulatory phosphorylation sites of TIF-IA (Schlosser et al., 2002; Zhao et al., 2003; Mayer et al., 2005; Hoppe et al., 2009), two of which are indeed controlled by the mammalian TOR pathway (Mayer et al., 2004). Rapamycin-induced inhibition of Pol I transcription correlates with both the inactivation of TIF-IA due to an altered phosphorylation pattern and its translocation to the cytoplasm resulting in impaired initiation-competent complex formation (Mayer et al., 2004). However, contradictory results were obtained by a study presenting UBF rather than hRRN3/TIF-IA as a downstream target of the mTOR pathway (Hannan et al., 2003).

Another TOR-dependent determinant of initiation-competent complex formation in yeast is the availability of Rrn3p. Contrary to TIF-IA, which is inactivated and exported from the nucleus upon rapamycin treatment (Mayer et al., 2004), the level of Rrn3p was recently reported to gradually decrease in likewise treated yeast cells due to the combination of proteasome-dependent degradation and a reduction in the neo-synthesis rate of this factor (Philippi et al., 2010). Consequently, a decrease in the level of Pol I-Rrn3p complex formation, in the association of Pol I with the rDNA locus and in 35S pre-rRNA synthesis was observable. The extent of the decrease could be diminished in all three cases by artificially stabilizing the level of Rrn3p in rapamycin-treated mutant cells (Philippi et al., 2010). The decrease in Pol I occupancy at the rDNA locus following rapamycin treatment could be further attenuated in mutant cells expressing an A43-Rrn3p fusion protein, thereby preventing not only the degradation of Rrn3p but also its dissociation from Pol I. Concomitantly, the decline in 35S pre-rRNA synthesis is also significantly retarded in these cells (Laferté et al., 2006).

INTRODUCTION

Although all these observations suggest distinct roles for Rrn3p-levels and for the phosphorylation status of both Rrn3p and Pol I in the regulation of Pol I-Rrn3p complex formation and thus Pol I transcription, little is known about the underlying regulatory mechansims. It is further unclear, to which extent these parameters contribute to the drastic decrease in ribosome production observed after TOR inactivation.

However, TOR inactivation affects ribosome biogenesis in yeast not only at the level of Pol I transcription initiation but also the elongation rate of the polymerase seems to be regulated in a growth-dependent manner. Recent findings demonstrated that in mutant cells characterized by the absence of Paf1C function, a complex which is shown to be positively involved in Pol I transcription elongation (Zhang et al., 2009), rRNA synthesis was affected to a significantly lower extent by impaired TOR signaling than likewise treated wild type cells (Zhang et al., 2010). This results suggests that Paf1C plays a TOR-dependent role in the modulation of rRNA production.

Besides Pol I transcription, TOR inactivation was also shown to specifically and rapidly down-regulate the RNA polymerase II-dependent transcription of ribosomal protein (RP) genes, which define a co-regulated cluster termed the RP regulon (Powers and Walter, 1999; Cardenas et al., 1999). Another regulon whose transcription by Pol II is similarly decreased following impaired TOR signaling is formed by the ribosome biogenesis (Ribi) genes coding for auxiliary ribosome biogenesis factors (Jorgensen et al., 2002, 2004). Consecutive analysis revealed several transcription regulators and transcription factors such as Sch9p, Sfp1p, Fhl1p and Ifh1p whose activity or binding to RP and Ribi gene promoters, respectively, is controlled by TOR signaling via alterations in their cellular localization or abundance (Jorgensen et al., 2004; Marion et al., 2004; Schawalder et al., 2004; Rudra et al., 2005). Since the factor Hmo1p was shown to bind both to ribosomal RNA and ribosomal protein genes in a rapamycin-sensitive manner (Hall et al., 2006; Berger et al., 2007), a function in coordinating Pol I and Pol II transcription in the context of ribosome biogenesis could be considered for this protein. The above effects on Pol II transcription will certainly contribute to the drop in ribosome production following inhibition of the TOR pathway, but to which extent remains to be further elucidated.

As mentioned before, kinase Tor1p displays a rapamycin-sensitive nuclear localization which is not only crucial for Pol I transcription but also for RNA polymerase III transcription (Li et al., 2006). The nuclear localization of TORC1 is apparently important for the phosphorylation and thus inactivation of the Pol III transcriptional repressor Maf1p (Wei et al., 2009), which is consistent with earlier reports showing that rapamycin treatment represses Pol III transcription in yeast (Zaragoza et al., 1998).

TOR inactivation obviously mediates the transcriptional down-regulation of all components required for ribosome biogenesis. However, the activity of RNA polymerase I seems to play a superior role in this process, since artificially stabilizing Pol I transcription in rapamycin-treated mutant cells expressing a constitutively initiation-competent version of Pol I attenuates the decrease in the level of both r-protein mRNAs and 5S rRNAs produced by Pol II and Pol III, respectively (Laferté et al., 2006).

In addition to transcription, general translation is also severely compromised upon TOR inactivation due to the impaired function of various translation factors as described (see section 2.3.2). 15 min of rapamycin treatment reduces the protein synthesis capacity of the cell by half (Barbet et al., 1996), however, the extent to which this process contributes to the down-regulation of ribosome biogenesis is again not well characterized.

Strikingly, it was shown that not only Pol I transcription is repressed following rapamycin treatment, but also 35S pre-rRNA processing is severely and very rapidly affected, thereby nearly abolishing the production of mature ribosomal RNAs (Powers and Walter, 1999). This effect could be derived from a direct TOR-dependent inactivation of ribosome biogenesis factors, from a rapid depletion of proteins indispensable for proper rRNA maturation due to a transcriptional and/or translational reduction in their expression level or from a combination of these processes. Indeed, decreasing levels of the ribosome biogenesis factors Nog1p and Nop7p were reported following TOR inactivation (Honma et al., 2006). Another example for TOR-mediated effects on RNA metabolism is the specific inhibition of splicing of r-protein mRNAs induced by amino acid starvation (Pleiss et al., 2007).

Finally, evidence exist suggesting that TOR signaling is involved in the control of pre-ribosomal transport processes. TOR inactivation leads to a rapid nucleolar entrapment of various ribosome biogenesis factors, thereby causing cessation of late rRNA maturation steps and defects in the nuclear-cytoplasmic translocation of pre-ribosomal particles (Honma et al., 2006; Vanrobays et al., 2008).

As mentioned before (see section 2.3.2), Sch9p, a genuine downstream target of TORC1, appears to play a central role in the regulation and coordination of ribosome biogenesis in response to environmental conditions. Besides its role in translation initiation, Sch9p was revealed both to be involved in the regulation of Pol I and Pol III transcription and to influence the expression of RP and Ribi genes. Notably, its influence on the transcriptional activity of RNA polymerase I is at least in part mediated by Rrn3p (Huber et al., 2009). However, since Sch9p is a non-essential protein and a constitutively active version of this protein confers only very slight resistance to rapamycin, its proposed role as a master regulator of protein synthesis should be at least considered questionable.

2.4 Objectives

Inactivation of the TOR pathway in eukaryotic cells appears to interfere with proper ribosome biogenesis at several different levels with varying stringency. Following rapamycin treatment or nutrient deprivation, both transcription by all three RNA polymerases and general translation as well as pre-rRNA processing are substantially down-regulated or exhibit severe defects, respectively. However, since these processes leading to mature ribosomes seem to be intimately linked and co-regulated, it is a challenging task to distinguish between primary and secondary effects.

INTRODUCTION

Therefore, the main objective of this thesis was to elucidate which of these processes is the dominant target of the TOR pathway provoking the immediate shut-off of mature rRNA production upon rapamycin treatment. To this end, RNA polymerase I transcription and rRNA synthesis were investigated shortly after TOR inactivation in yeast, thereby trying to uncouple these two processes in order to analyze their response to rapamycin separately.

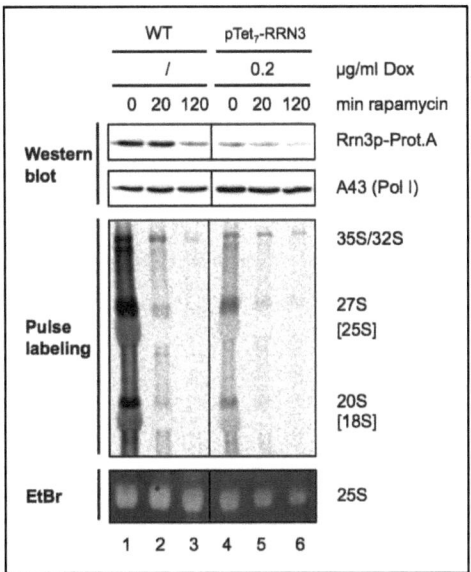

Figure 7. Reduced levels of Rrn3p upon TOR inactivation does not explain the complete shut-off of rRNA synthesis.
Although the level of Rrn3p is lowered to an equal extent in a mutant strain and a corresponding wild type strain either by artificially reducing the amount of the protein or by treating the cells with rapamycin for 120 min, respectively, the observed effects on rRNA synthesis in the two strains differ substantially (compare lane 3 with lane 4). Just additional rapamycin treatment for the same time decreases rRNA production in the mutant strain to a comparable extent (compare lane 3 with lane 6). Obviously, reduced levels of Rrn3p contribute to the down-regulation of both Pol I transcription and rRNA synthesis, however, there must be additional mechanisms which mediate the complete shut-off of mature rRNA production in yeast cells upon TOR inactivation. [from (Philippi et al., 2010)]

By artificially down-regulating the level of Rrn3p with doxycycline in a mutant strain expressing this protein under the control of a sevenfold tetracycline-regulable promoter, overall rRNA synthesis could also be decreased to some extent compared to an untreated wild type strain (Figure 7, compare lane 4 with lane 1). However, when the same Rrn3p-level is generated in the wild type strain by rapamycin treatment for 120 min, the production of ribosomal RNA is nearly completely abolished (Figure 7, compare lane 4 with lane 3). In both strains, significant pre-rRNA processing defects are already observable after 20 min of rapamycin treatment, although the respective level of Rrn3p is yet unaltered (Figure 7, compare lane 1 with lane 2 and lane 4 with lane 5). These findings already suggest that the drastic down-regulation of mature rRNA production observed after TOR inactivation is not only accomplished by decreasing the level of

INTRODUCTION

Rrn3p and consequently RNA polymerase I transcription, but also by interfering with pre-rRNA processing.

Furthermore, since the level of Rrn3p appears to play a distinct role in the regulation of Pol I-Rrn3p complex formation and thus Pol I transcription and ribosome biogenesis, the impact of Rrn3p overexpression on rRNA synthesis and yeast cell growth was investigated in more detail. Additionally, Rrn3p was purified from yeast cell extracts in order to identify phosphorylation sites of this phosphoprotein which presumably play an important role in the process of Pol I-Rrn3p complex formation. In parallel, co-purifying proteins were analyzed closely, since they constitute putative interaction partners of Rrn3p which might also be involved in the regulation of Rrn3p function.

3 RESULTS

3.1 Effects of TOR inactivation on RNA polymerase I transcription, rRNA production, and yeast cell growth

3.1.1 Proteasome-dependent reduction of Rrn3p-levels in growth-arrested yeast cells

It has been previously shown that there is a concomitant down-regulation of rRNA synthesis and Pol I-Rrn3p complex formation in stationary yeast cells (Milkereit and Tschochner, 1998). Furthermore, when cells are treated with rapamycin, a drug which mimics conditions of nutrient starvation by inactivating the kinase Tor1p, or with the protein synthesis inhibitor cycloheximide, both a reduction of Pol I-Rrn3p complexes and pre-rRNA synthesis could be observed (Yuan et al., 2002; Cavanaugh et al., 2002; Hirschler-Laszkiewicz et al., 2003; Mayer et al., 2004; Claypool et al., 2004).

This reduction of Pol I-Rrn3p complexes in yeast cells appeared to correlate with decreasing amounts of Rrn3p in whole cell extracts (WCE) (Philippi, 2008). To confirm these results and to elucidate further the general mechanism underlying the reduction of Rrn3p-levels, exponentially growing yeast cells of the strain RRN3-Prot.A were treated with either rapamycin (RAPA) or cycloheximide (CHX). Indeed, the level of Rrn3p dropped continuously in both cases (Figure 8, upper panel). After 120 min of rapamycin or cycloheximide treatment, the level of this factor decreased below 20% of the initial amount. In contrast, Pol I-levels remained rather stable within this time frame (Figure 8, lower panel).

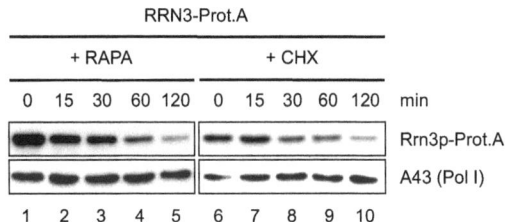

Figure 8. Reduction of Rrn3p-levels in growth arrested yeast cells.
Yeast strain Y2183 expressing a chromosomally Prot.A-tagged Rrn3p was grown in YPD at 30°C to mid-log phase (OD_{600} ~ 0,4) before cells were either treated with 200 ng/ml of rapamycin (RAPA) or with 100 µg/ml of cycloheximide (CHX). At the time points indicated, cells were collected and lyzed. Same amounts of WCE (20 µg) were analyzed by Western blotting using antibodies directed against the Prot.A-tag of Rrn3p and the Pol I subunit A43, respectively.

These observations demonstrate that inactivation of both TOR pathway and protein synthesis results in a reduction in the level of Pol I transcription initiation factor Rrn3p. The quick down-regulation of Rrn3p-levels in growth arrested yeast cells was recently suggested to be dependent on the proteolytic activity of the proteasome (Philippi et al., 2010). In the temperature-sensitive mutant strain cim3-1-RRN3-TAP, which is defective in *CIM3/SUG1*, a gene

RESULTS

coding for an essential subunit of the regulatory particle of the 26S proteasome (Rubin et al., 1996; Gerlinger et al., 1997), Rrn3p-levels are stable when TOR signaling is impaired upon amino acid depletion at the restrictive temperature (Philippi et al., 2010). Since ubiquitylation targets a protein for destruction by the proteasome system, the ubiquitylation status of Rrn3p was analyzed before and after inhibition of the TOR pathway. It was previously reported that (poly)ubiquitylated proteins can be specifically enriched by their affinity to the immobilized (poly)ubiquitin-binding protein Dsk2p (Funakoshi et al., 2002). Whole cell extracts generated from the strain pNOP1-RRN3-Prot.A [pGAL1-Myc$_3$-UBI4 [G76A]] expressing Prot.A-tagged Rrn3p and Myc$_3$-tagged ubiquitin before and after 10 min of rapamycin treatment were incubated with GST (Figure 9A, lanes 1-7 and Figure 9B, lanes 1-8) or GST-Dsk2p (Figure 9A, lanes 8-14 and Figure 9B, lanes 9-16) fusion protein bound to glutathione sepharose. After several wash steps, GST-baits and attached proteins were eluted and subjected to Western blot analysis.

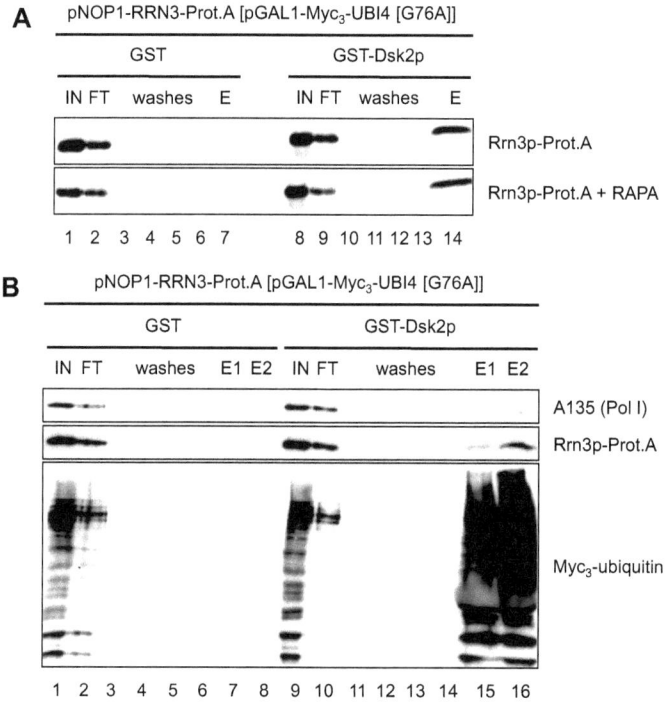

RESULTS

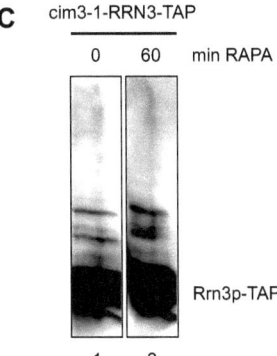

Figure 9. Specific enrichment of (poly)ubiquitylated Rrn3p by (poly)ubiquitin-binding protein Dsk2p.
(A) Yeast strain Y2181 expressing Prot.A-tagged Rrn3p from a plasmid under the control of the *NOP1* promoter was grown in YPD at 30°C to mid-log phase before half of the cells were treated with 200 ng/ml of rapamycin (RAPA) for 10 min. Cells of rapamycin treated and untreated cultures were collected and lyzed. Same amounts of WCE (6 mg) were incubated with either recombinant GST or recombinant GST-Dsk2p immobilized on 50 µl of glutathione sepharose. After washing, proteins bound to the beads were eluted with SDS sample buffer. 1% of input (IN) and flow through (FT), 0.5% of the wash steps (washes) and 50% of the eluate (E) were analyzed by Western blotting using antibodies directed against the Prot.A-tag of Rrn3p.
(B) The same experiment as described in Figure 9A, except for growing yeast strain Y2181 in YPG at 30°C to induce expression of Myc$_3$-tagged ubiquitin under the control of the *GAL1* promoter and without rapamycin treatment. Retention of Pol I subunit A135 on the GST-Dsk2p-beads was analyzed by Western blotting using antibodies directed against A135. Ubiquitylation was analyzed using antibodies directed against the Myc$_3$-tag of ubiquitin.
(C) The proteasome temperature-sensitive mutant strain Y652 expressing chromosomally TAP-tagged Rrn3p was grown in YPD at 24°C to mid-log phase before cells were shifted to 37°C for 2 h. Half of the culture was collected and lyzed, whereas incubation at 37°C was continued for 1 h with the remainder of the cells in the presence of 200 ng/ml rapamycin prior to harvest and lysis. Same amounts of WCE (20 mg) were incubated with either recombinant GST-Dsk2p or recombinant GST immobilized on 50 µl of glutathione sepharose. After washing, proteins bound to the beads were eluted with SDS sample buffer. 90% of the eluate (E) was analyzed by Western blotting using antibodies directed against the Prot.A-tag of Rrn3p. At longer exposure times, Rrn3p species of higher molecular weight were detected (polyubiquitylated Rrn3p).

Rrn3p was specifically enriched by GST-Dsk2p but not GST alone (Figure 9A and Figure 9B, central panel). Interestingly, rapamycin treatment does not further induce ubiquitylation of Rrn3p, since this factor is apparently constitutively ubiquitylated and degraded. Therefore, down-regulation of Rrn3p-levels upon TOR inactivation must be achieved by a reduction in the neo-synthesis of this protein. Indeed, *RRN3* mRNA-levels are reduced to 30% after 20 min of rapamycin treatment (Philippi et al., 2010), which is in good agreement with previous transcriptome analyses (Huang et al., 2004). In addition, general translation was also reported to be decreased to 50% following 15 min of rapamycin treatment, thus providing further evidence for reduced neo-synthesis of Rrn3p (Barbet et al., 1996).

Notably, Rrn3p, which is detected in the eluates, seems to represent the monoubiquitylated form, since it migrates with a slightly lower mobility in SDS-polyacrylamide gel electrophoresis (SDS-PAGE) compared to Rrn3p in the input and flow through fractions (Figure 9A, compare lanes 8 and 9 with 14). It seems that polyubiquitylated species of Rrn3p are highly unstable and could therefore just be detected when Rrn3p was enriched by GST-Dsk2p in the proteasome-deficient strain cim3-1-RRN3-TAP (Figure 9C). Whole cell extracts of this strain, cultivated at the restrictive temperature, contain additionally higher migrating species of Rrn3p which likely represent polyubiquitylated forms. Interestingly, preceding rapamycin treatment for 1 h does not increase the amount of both mono- and polyubiquitylated Rrn3p underlining again the constitutive degradation of this protein.

In contrast, Pol I subunit A135 is not among the (poly)ubiquitylated protein species which are specifically enriched by GST-Dsk2p (Figure 9B, upper and lower panel) suggesting that A135 is not a target of ubiquitylation under these conditions.

3.1.2 Level of Rrn3p influences Pol I-Rrn3p complex formation, Pol I recruitment to the rDNA, and yeast cell growth but not the rDNA copy number

Apparently, a connection between Rrn3p-levels and cell growth exists, since previous work demonstrated that a decline in the abundance of Rrn3p is accompanied with a reduction in the growth rate of yeast cells (Philippi, 2008). To analyze to what extent artificially altered Rrn3p-levels influence Pol I transcription and cell growth, the yeast mutant strains pTet$_7$-RRN3-Prot.A and pTet$_7$-RRN3-Prot.A-A43-HA$_3$ were used, in which the expression levels of Rrn3p could be adjusted by the drug doxycycline (Dox). In these strains, an endogenous deletion of the *RRN3* gene is rescued by a plasmid expressing RRN3-Prot.A fusion protein under the control of a sevenfold tetracycline-regulable promoter (Garí et al., 1997; Philippi, 2008). Additionally, in the latter, Pol I subunit A43, which interacts with Rrn3p in the course of Pol I-Rrn3p complex formation (Peyroche et al., 2000), is endogenously expressed with a C-terminal HA$_3$-tag important for subsequent chromatin immunoprecipitation (ChIP) analysis. Without doxycycline in the medium, a derivative of tetracycline, Rrn3p is heavily overexpressed. By adding increasing amounts of doxycycline to the medium, Rrn3p expression could artificially be down-regulated (Figure 11, upper panel).

The mutant strain pTet$_7$-RRN3-Prot.A grows best when wild type expression levels of Rrn3p are adjusted by adding 0.1 µg/ml doxycycline (Figure 10), a concentration established by preliminary tests and experiments with the strain pTet$_7$-RRN3-Prot.A-A43-HA$_3$ (Figure 11A, lane 5). Interestingly, both overexpression (see section 3.2) and down-regulated levels of Rrn3p cause a reduced growth rate of the mutant strain pTet$_7$-RRN3-Prot.A (Figure 10) (Philippi, 2008).

The strain pRRN3-Prot.A-A43-HA$_3$, which is identical to the corresponding mutant strain pTet$_7$-RRN3-Prot.A-A43-HA$_3$ except for expressing RRN3-Prot.A fusion protein under the control of the endogenous promoter, served as an appropriate doxycycline-insensitive control.

RESULTS

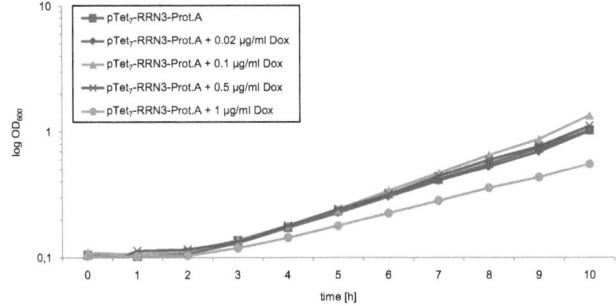

Figure 10. Influence of different expression levels of Rrn3p on yeast cell growth.
Yeast strain Y667 expressing Prot.A-tagged Rrn3p only from a plasmid under the control of a sevenfold tetracycline-regulable promoter was grown in YPD at 30°C either in the absence or in the presence of the indicated doxycycline concentrations (Dox). The cultures were diluted to OD_{600} = 0.1. The OD_{600} of the cultures was monitored hourly over the entire 10 h time course of the experiment.

As mentioned before, adding 0.1 µg/ml of doxycycline to the medium results in a nearly wild type expression level of Rrn3p in the strain pTet$_7$-RRN3-Prot.A-A43-HA$_3$ when compared to the control strain (Figure 11A, compare lane 5 with lane 1 and lane 2). In contrast to the mutant strain pTet$_7$-RRN3-Prot.A, in this mutant wild type Rrn3p-levels lead to an intermediate growth rate, whereas raised or depressed amounts of this factor provoke higher or lower growth rates, respectively (data not shown). The most likely explanation for this phenomenon is that overexpression of Rrn3p also leads to a growth defect probably due to excessive complex formation (see section 3.2). The HA$_3$-tag of Pol I subunit A43 apparently interferes with this enhanced complex formation, thereby repressing this growth phenotype in the strain pTet$_7$-RRN3-Prot.A-A43-HA$_3$ under Rrn3p overexpression conditions.

Notably, even treatment with a high dose of doxycycline (1 µg/ml) does not inhibit growth of both mutant strains completely (Figure 10) (data not shown), although Rrn3p-levels are down-regulated to at least the same degree as those after 2 h of rapamycin treatment (compare Figure 8, lane 1 and 5 with Figure 11A, lane 5 and 7).

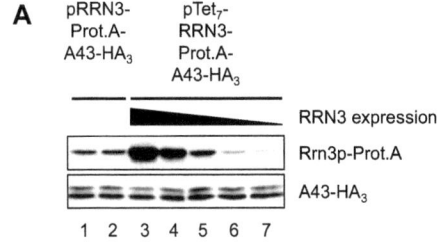

RESULTS

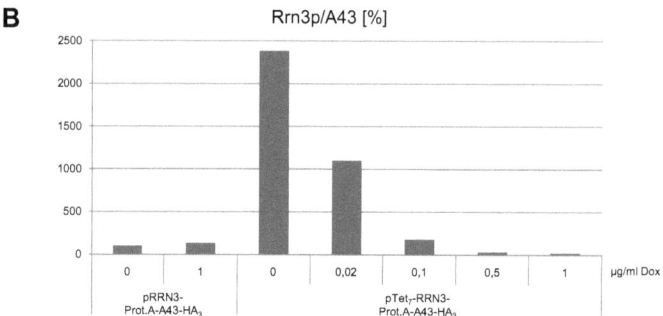

Figure 11. Doxycycline-dependent expression of Rrn3p in the strain pTet$_7$-RRN3-Prot.A-A43-HA$_3$.
(A) Yeast strains Y2186 and Y2185, both expressing HA$_3$-tagged Pol I subunit A43 and Prot.A-tagged Rrn3p under the control of the endogenous or the sevenfold tetracycline-regulable promoter, respectively, were grown in YPD at 30°C in the presence of the indicated doxycycline concentrations to mid-log phase before cells were collected and lyzed. Same amounts of WCE (20 µg) were analyzed by Western blotting using antibodies directed against the Prot.A-tag of Rrn3p and the HA$_3$-tag of the Pol I subunit A43, respectively.
(B) Quantitative analysis of the signals presented in Figure 11A was performed with the Multi Gauge software (Fujifilm). Rrn3p-Prot.A signals were normalized to the signals of A43-HA$_3$ which served as a loading control. The level of Rrn3p of the control strain grown without doxycycline was arbitrarily set to 100%.

This suggests, that the effects of rapamycin treatment are not solely mediated by the decrease in the expression levels of Rrn3p. There must be other mechanisms which contribute to the impairment of cell growth when TOR signaling is inhibited. Furthermore, it became obvious that very small amounts of Rrn3p are sufficient to drive growth.

Next, the question arose as to whether different Rrn3p-levels cause an alteration in Pol I-Rrn3p complex formation in the mutant strain pTet$_7$-RRN3-Prot.A. To this end, gel filtration experiments were performed with WCE generated from this strain grown either in the absence or in the presence of 1 µg/ml doxycycline. It has been previously shown that Rrn3p is present in three different forms in whole cell extracts of exponentially growing yeast cells (Bier et al., 2004). 21% of the recovered Rrn3p elutes with the void volume of the column, about 24% co-migrates with Pol I, whereas the largest fraction containing 55% of Rrn3p migrates with the molecular mass of monomeric Rrn3p. When the cellular level of Rrn3p is altered in a specific way, then the extent of the void fraction, the Pol I-complexed fraction and the monomeric fraction of Rrn3p is altered in the same manner (Figure 12). Apparently, there is a cellular mechanism maintaining a fixed ratio between free and complexed Rrn3p. This means that enhanced levels of Rrn3p lead to a higher amount of Pol I-Rrn3p complexes, whereas a reduced Rrn3p expression has the opposite effect, demonstrating the correlation between Rrn3p-levels and the number of Pol I-Rrn3p complexes. Notably, changes in the expression level of Rrn3p does not significantly influence the elution profile of both this protein and RNA polymerase I in the gel filtration experiment.

RESULTS

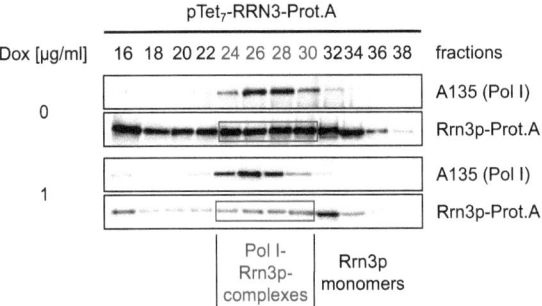

Figure 12. Overexpression of Rrn3p results in increased amounts of Pol I-Rrn3p complexes.
Yeast strain Y667 expressing Prot.A-tagged Rrn3p under the control of the sevenfold tetracycline-regulable promoter was grown in YPAD at 30°C in the absence or in the presence of 1 μg/ml doxycycline to mid-log phase before cells were collected and lyzed. Same amounts of WCE (900 μg) were separated on a Superose 6® column in a buffer containing 1.5 M potassium acetate. 50% (250 μl) of every second fraction were TCA precipitated and analyzed by Western blotting using antibodies directed against the Prot.A-tag of Rrn3p and the Pol I subunit A135, respectively.

The observation that the amount of Pol I-Rrn3p complexes could be artificially increased and decreased by changing the expression level of Rrn3p raised the question whether the amount of Pol I being recruited to the rDNA could be influenced as well. To get insight into this issue, Pol I-ChIP experiments were performed with the mutant strain pTet$_7$-RRN3-Prot.A-A43-HA$_3$. Cells were grown in YPD at 30°C to mid-log phase in the absence or in the presence of different doxycycline concentrations (0.02 μg/ml; 0.1 μg/ml; 0.5 μg/ml; 1 μg/ml) before cells were crosslinked with 1% formaldehyde for 15 min, harvested, lyzed and sonified. The HA$_3$-tagged Pol I subunit A43 was immunoprecipitated from the chromatin extracts. After DNA isolation, the relative amounts of specific DNA fragments co-purifying with the protein were measured in triplicate real-time PCR reactions. RNA Pol I crosslinking to the 35S rRNA gene promoter and to two regions coding for the 18S rRNA and the 25S rRNA was examined. The 5S rRNA-coding region was included as an internal control.

Doxycycline-insensitive strain pRRN3-Prot.A-A43-HA$_3$, either untreated or treated with 1 μg/ml of doxycycline, was also included in the experiment.

Indeed, regarding the mutant strain, a stepwise reduction of approximately 60% of both promoter DNA and DNA fragments of the 35S rRNA-coding sequence co-precipitating with Pol I could be noticed when Rrn3p-levels and concomitantly Pol I-Rrn3p complexes are diminished (Figure 13, right hand side). In contrast, no change in the pattern of Pol I recruitment to the rDNA could be detected in the doxycycline-insensitive control strain (Figure 13, left hand side). Furthermore, it seems that the decrease in promoter occupancy of Pol I is slightly stronger than that in the transcribed region of the rDNA. This could indicate that promoter clearance is a limiting step in cells with wild type and enhanced Rrn3p-levels. However, despite a more than 20 fold increase in the amount of Rrn3p in the mutant cells without doxycycline treatment, no increase in Pol I occupancy at the rDNA locus could be detected in this ChIP analysis. Nevertheless, the existence of such an increase could not be excluded due to possible

RESULTS

limitations of the chromatin immunoprecipitation method in this special case (see section 3.2.3). For example, both one and five Pol I molecules crosslinked to a specific DNA fragment will just result in the precipitation of this single fragment, concealing a fivefold increase in the Pol I loading on this DNA. In contrast, a reduction of Pol I molecules is detectable without limitations. The Miller chromatin spreading technique could contribute to the resolution of this open question, since this method can visualize Pol I molecules on individual rRNA genes by electron microscopy (Miller and Beatty, 1969).

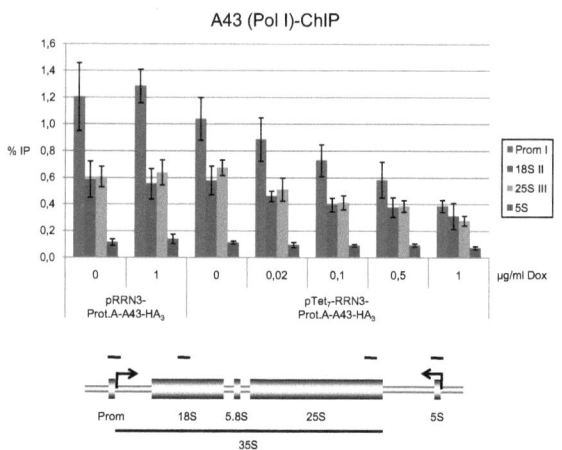

Figure 13. Decreasing levels of Rrn3p correlate with a reduced rate of Pol I recruitment to the 35S rDNA.
Yeast strains Y2186 and Y2185, both expressing HA_3-tagged Pol I subunit A43 and Prot.A-tagged Rrn3p under the control of the endogenous or the sevenfold tetracycline-regulable promoter, respectively, were grown in YPD at 30°C in the presence of the indicated doxycycline concentrations to mid-log phase before cells were crosslinked with 1% formaldehyde, harvested, lyzed, and sonified. The HA_3-tagged Pol I subunit A43 was immunoprecipitated from the chromatin extracts. After DNA isolation, the relative amounts of specific DNA fragments co-purifying with the protein were measured in triplicate real-time PCR reactions using primers specific for the rDNA promoter (Prom I), the 18S (18S II) and 25S (25S III) rRNA-coding region as well as for the 5S rRNA gene (5S), which served as an internal control. The data represent the mean of at least three independent ChIP experiments. The positions of the amplified rDNA regions are indicated in a sketch on the bottom of the figure.

It has also been shown that doxycycline-dependent reduction of Pol I recruitment to the rDNA in the mutant strain pTet$_7$-RRN3-Prot.A leads to a down-regulation of Pol I transcription without any pre-rRNA processing defects (Philippi et al., 2010). In summary, these results demonstrate a correlation between Rrn3p-levels, the formation of Pol I-Rrn3p complexes, recruitment of Pol I to the rRNA genes, rRNA gene transcription and the growth rate in yeast. It is not clear, however, whether alterations in the level of Rrn3p induce variations in the number of rDNA repeat units.

As already mentioned, the rRNA genes in *Saccharomyces cerevisiae* consist of 100-140 transcription units, arranged in a tandem array on chromosome XII (Petes, 1979). In a single cell, only about half of the rDNA repeats are actively transcribed, whereas the other half of the population is transcriptionally inactive (Dammann et al., 1993). The number of the repeats is dynamic and can vary due to unequal meiotic and mitotic recombination events (Warner, 1989).

RESULTS

To investigate whether an artificial change in the level of the transcription factor Rrn3p leads to an alteration in the number of rDNA copies of the yeast cell, the content of rDNA in the corresponding input fractions of the preceding ChIP experiment were analyzed by real-time PCR reactions using primers specific for the 25S rRNA-coding region. The raw data were normalized to the content of DNA in the samples determined by using primers specific for the single copy gene locus *NOC1*. Yeast strains NOY1071 and NOY1064, carrying approximately 25 or 190 rDNA repeats, respectively, served as internal controls (Cioci et al., 2003). Notably, no significant change in the number of the rDNA copies could be detected (Figure 14).

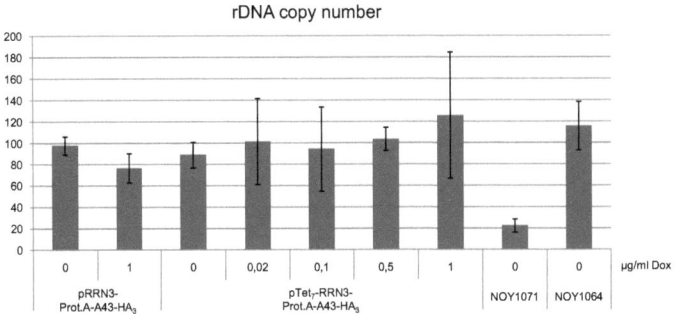

Figure 14. Variations in Rrn3p-levels do not result in an alteration in the rDNA copy number.
Yeast strains Y2186 and Y2185, both expressing HA$_3$-tagged Pol I subunit A43 and Prot.A-tagged Rrn3p under the control of the endogenous or the sevenfold tetracycline-regulable promoter, respectively, and yeast strains Y353 and Y352, carrying approximately 25 or 190 rDNA repeats, respectively, were analyzed as described for Figure 13. After DNA isolation, the content of rDNA was measured in triplicate real-time PCR reactions using primers specific for the 25S rRNA-coding region and normalized to the content of DNA using primers specific for the single copy gene locus *NOC1*.

Furthermore, it may be possible that changes in the expression level of Rrn3p cause an alteration in the ratio of actively transcribed to transcriptionally inactive rRNA gene repeats. To address this question, several psoralen crosslinking experiments were performed, however, due to technical problems, no adequate statements elucidating this issue could be made.

Since rapamycin treatment both results in down-regulation of Rrn3p-levels and growth inhibition of yeast cells, the question arose whether maintaining certain levels of Rrn3p in yeast cells could withstand or at least attenuate growth inhibition of the cells. To this end, growth kinetics of yeast strains RRN3-Prot.A and pTet$_7$-RRN3-Prot.A were monitored. The latter was grown either in the absence or in the presence of 1 µg/ml doxycycline. After 6 h, rapamycin was added to a final concentration of 200 ng/ml to the moiety of each culture. As previously described, different Rrn3p-levels result in different growth rates of the yeast cells. However, when rapamycin is added to the cultures, all three stop growing within a time frame of approximately 1 h. Rapamycin-induced growth arrest occurred with the same kinetics in all three cultures, no matter whether Rrn3p was overexpressed or scarce (Figure 15).

RESULTS

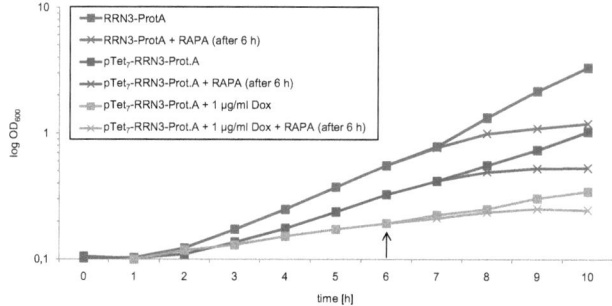

Figure 15. Expression levels of Rrn3p do not influence the kinetics of growth inhibition upon TOR inactivation.
Yeast strain Y2183 expressing Prot.A-tagged Rrn3p and yeast strain Y667 expressing Prot.A-tagged Rrn3p only from a plasmid under the control of a sevenfold tetracycline-regulable promoter were grown in YPD at 30°C either in the absence or in the presence of 1 µg/ml doxycycline (note that strain Y2183 is not the exact isogenic wild type strain to strain Y667). The cultures were diluted to OD_{600} = 0.1. After 6 h, half of the cultures were withdrawn and rapamycin was added to a final concentration of 200 ng/ml (indicated by an arrow). The OD_{600} of the cultures was monitored hourly over the entire 10 h time course of the experiment.

Therefore, the simple reduction of Rrn3p-levels and consequently the decreased association of the Pol I machinery with the rDNA locus do not fully explain the drastic negative effects on ribosome neo-production (Figure 7, compare lane 4 with lane 3) and on cellular growth (Figure 15) upon TOR inhibition. It rather seems that the severe and immediate effect on pre-rRNA processing, predominantly affecting the production of the intermediate 27S and 20S RNAs and the mature 25S and 18S rRNAs, is the dominant negative effect of short-term TOR inactivation (Figure 7, lane 2). Although the corresponding 35S pre-rRNA-levels after 20 min of rapamycin treatment are comparable to those observed in untreated cells, it is not clear whether Pol I transcription is indeed unaffected (Figure 7, compare lane 2 with lane 1 and lane 5 with lane 4). Since 35S pre-rRNA-levels are the result of the co-action of pre-rRNA production, pre-rRNA processing and pre-rRNA degradation, a straightforward interpretation of the pulse labeling data is quite difficult. To this end, the effects of impaired TOR signaling on RNA Pol I transcription are investigated more accurately.

3.1.3 RNA polymerase I transcription is not affected at early stages of TOR inactivation in yeast cells

Pol I transcription appears not only to be regulated by the abundance of the transcription factor Rrn3p. It was also reported that modulations of the posttranslational modification pattern of both Rrn3p and RNA polymerase I influence Pol I transcription, for example by mediating Pol I-Rrn3p complex formation or elongation rate (Paule et al., 1984; Fath et al., 2001, 2004; Cavanaugh et al., 2002; Yuan et al., 2002; Hirschler-Laszkiewicz et al., 2003; Zhao et al., 2003; Mayer et al., 2004, 2005; Gerber et al., 2008). Therefore, Pol I transcription could still be affected more drastically upon TOR inactivation than the moderate down-regulation of Rrn3p-levels and

RESULTS

35S pre-rRNA-levels suggests. Hence, RNA polymerase I association with the 35S rRNA gene before and after various time points of rapamycin addition was investigated. ChIP experiments were carried out using the yeast strain RRN3-TAP-A43-HA$_3$. Exponentially growing cells were incubated either in the absence or in the presence of rapamycin. Immediately before as well as 15 min, 30 min and 60 min after rapamycin addition, two samples were withdrawn of both the untreated and the treated culture and either fixed with 1% formaldehyde for 15 min to perform ChIP experiments or used for the preparation of whole cell extracts and subsequent Western blot analysis to monitor the corresponding Rrn3p-levels. Again, RNA Pol I crosslinking to the 35S rRNA gene promoter, to two regions coding for the 18S rRNA and the 25S rRNA and to the 5S rRNA-coding region, which served as an internal control, was examined.

As expected, when yeast cells were not treated with rapamycin during exponential growth in a control experiment, no significant changes both in Rrn3p-levels and Pol I occupancy at the rDNA locus could be observed (Figure 16A and 16B). No down-regulation, but rather a slight increase in both levels is detectable within this time frame.

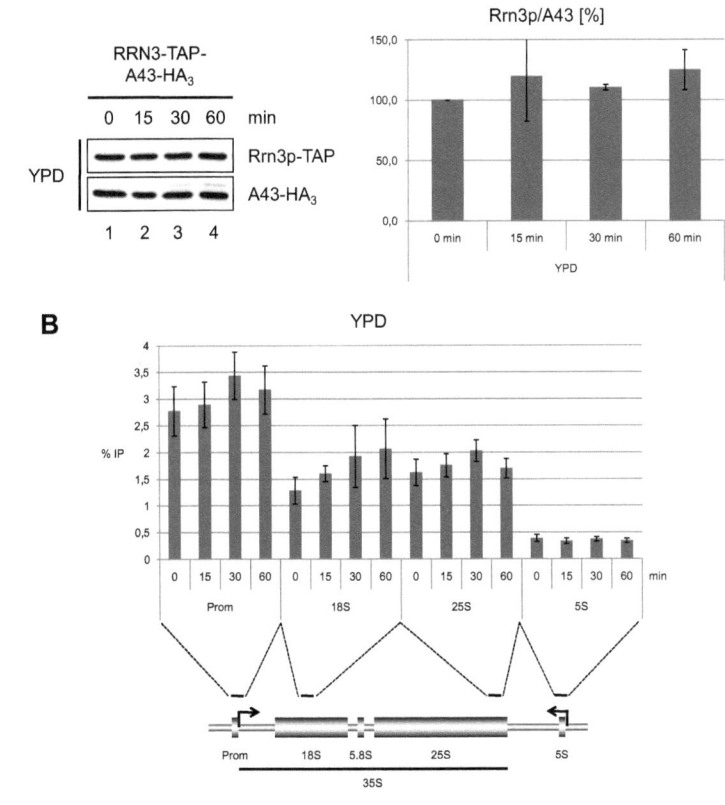

RESULTS

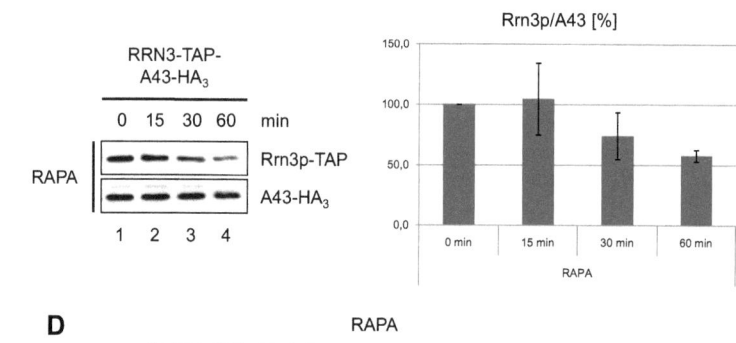

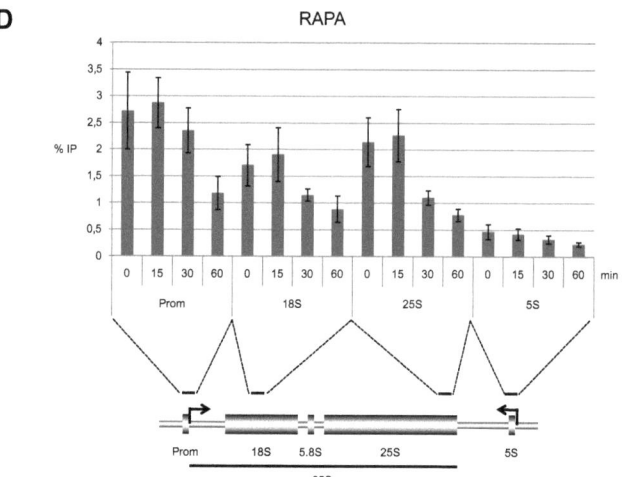

Figure 16. Endogenous Rrn3p-levels and Pol I association with 35S rRNA genes are unaltered after 15 min of rapamycin treatment, but decrease after 30 min and 60 min of drug treatment.
Yeast strain Y658 expressing HA$_3$-tagged Pol I subunit A43 and TAP-tagged Rrn3p was grown in YPD at 30°C to mid-log phase before the culture was split in two parts and further cultivated in YPD either in the absence or in the presence of rapamycin (200 ng/ml). At the time points indicated, two samples were withdrawn of both the untreated and the treated culture and cells were either collected and lyzed for subsequent Western blot analysis **(A/C)** or treated with 1% formaldehyde at 30°C for 15 min, harvested, lyzed, and sonified for subsequent ChIP experiments **(B/D)**.
(A/C) Same amounts of WCE (10 µg) were analyzed by Western blotting using antibodies directed against the TAP-tag of Rrn3p and the HA$_3$-tag of Pol I subunit A43, respectively. Quantitative analysis of the signals was performed with the Multi Gauge software (Fujifilm). Rrn3p-TAP signals were normalized to the signals of A43-HA$_3$, which served as a loading control. The level of Rrn3p of the first time point was arbitrarily set to 100%.
(B/D) The HA$_3$-tagged Pol I subunit A43 was immunoprecipitated from the chromatin extracts. After DNA isolation, the relative amounts of specific DNA fragments co-purifying with the protein were measured in triplicate real-time PCR reactions using primers specific for the rDNA promoter (Prom I), the 18S (18S II) and 25S (25S III) rRNA-coding region as well as for the 5S rRNA gene (5S), which served as an internal control. The data represent the mean of at least three independent ChIP experiments. The positions of the amplified rDNA regions are indicated in a sketch on the bottom of the figure.

However, as already shown in Figure 8, when cells were treated with rapamycin, a reduction of the level of Rrn3p could be stated. But also a decline of Pol I recruitment to the rDNA locus occurred, which correlates well with the down-regulation of Rrn3p-levels. After 30 min and 60 min of treatment, both the level of Rrn3p and the amount of the rRNA gene fragments co-

RESULTS

precipitating with RNA Pol I subunit A43 were decreased to about half of the level before drug application (Figure 16C and 16D), the latter of which is consistent with previous studies (Claypool et al., 2004). Regarding the similar kinetics of both declines, it appears that just the level of Rrn3p determines the association of Pol I with the rDNA under these conditions.

Remarkably, at early stages of TOR inactivation due to 15 min of rapamycin treatment, no reduction in both Rrn3p-levels and 35S rRNA gene fragments co-precipitating with RNA Pol I subunit A43 could be detected (Figure 16C and 16D), but strong pre-RNA processing defects are already observable after 20 min of treatment (Figure 7, lane 2).

Despite the evidence of Pol I molecules still being engaged on the rRNA genes at an early time point of impaired TOR signaling, it is not clear yet whether 35S pre-rRNA is actually produced. The possibility remained that these polymerase molecules are transcriptionally inactive and stalled on the rDNA resulting in no neo-production of 35S pre-rRNA. To elucidate the integrity of the Pol I transcription machinery at the rRNA genes after 15 min of rapamycin treatment, chromatin endogenous cleavage (ChEC) experiments were performed (Schmid et al., 2004). This method allows to precisely map association of proteins with the DNA within large genomic regions and to get a quantitative estimate which percentage of this DNA is decorated with the respective protein. Specifically the question whether the RNA Pol I molecules can still leave the transcribed region after TOR inactivation and inhibition of transcription (re)initiation was addressed. To this end, the two isogenic yeast strains RRN3-A43-MNase-HA$_3$ and rrn3-8-A43-MNase-HA$_3$ were used, both expressing Pol I subunit A43 from its genomic locus as a fusion protein with a C-terminally HA$_3$-tagged micrococcus nuclease (MNase). The latter of the strains, however, carries a temperature-sensitive allele of *RRN3* (*rrn3-8*). It is defective in RNA Pol I transcription (re)initiation upon a shift to the restrictive temperature (Cadwell et al., 1997). The outline of the experiment is illustrated in Figure 17A. The two strains were grown in YPD at 24°C to mid-log phase before the cultures were split in two parts and rapamycin was added to one half of the culture. Cells were cultivated for additional 15 min, then one half of each culture was withdrawn and subjected to formaldehyde crosslinking for subsequent ChEC analysis. The remainder of the cultures was shifted to 37°C and incubated for another 90 min before the cells were also treated with formaldehyde and analyzed by ChEC.

After formaldehyde crosslinking, cells were harvested and nuclei were prepared before calcium was added to activate the endonuclease activity of the fusion proteins. Thus, the DNA was cut in the proximity of the protein-binding site. The reaction was stopped by the addition of the calcium chelating reagent EDTA, total DNA was isolated, and linearized with the restriction endonuclease XcmI. The restriction fragments were separated by agarose gel electrophoresis prior to being transferred to a membrane. Indirect endlabeling analysis allows the precise localization of the cleavage sites mediated by the MNase fusion proteins.

Consistent with previous results, a characteristic cleavage pattern of the A43-MNase fusion protein could be observed with strong cleavage events at the promoter of the 35S rRNA gene

RESULTS

and throughout the entire Pol I-transcribed sequence (Merz et al., 2008) (Figure 17B and 17C, lanes 3-5, 8-10, 13-15 and 18-20).

Interestingly, no significant changes in the A43-MNase-mediated cleavage pattern before and after 15 min of rapamycin treatment could be noticed in cells grown at 24°C, regardless of expressing the wild type or the mutant allele of *RRN3* (Figure 17B and 17C, compare lanes 1-5 with lanes 6-10). This is in good agreement with the observation that Rrn3p-levels and Pol I association with the rDNA locus is unaltered before and after 15 min of TOR inactivation (Figure 16C and 16D). After the temperature shift of strain RRN3-A43-MNase-HA$_3$ to 37°C for 90 min, the extent of the cleavages of MNase-tagged Pol I at the rDNA promoter region is reduced in chromatin from cells treated with rapamycin compared to chromatin from untreated cells (Figure 17B, compare lanes 13-15 with lanes 18-20 and Figure 17D). This again correlates well with the decrease in both Rrn3p-levels and Pol I association with the rRNA gene promoter and the transcribed region after prolonged rapamycin treatment observed in the ChIP experiment shown in Figure 16C and 16D. Notably, after more than 100 min, still substantial amounts of Pol I molecules are engaged in rDNA transcription.

In contrast, when cells of the mutant strain rrn3-8-A43-MNase-HA$_3$ are shifted to 37°C for 90 min, the Pol I-MNase-mediated cleavage pattern is drastically reduced in chromatin from rapamycin treated and untreated cells (Figure 17C, lanes 11-20 and Figure 17D). Since no additional Pol I molecules can (re)initiate transcription due to the temperature shift, possible cleavage events could only derive from Pol I molecules still being associated to the rDNA gene. The almost complete loss of cleavage events suggests that in both cases the engaged Pol I molecules were able to finish their transcription cycle. This result led to the conclusion that TOR inactivation does not immobilize RNA Pol I molecules on the rDNA template.

Similar results were obtained, when these strains were depleted of essential amino acids for 2 h prior to a shift to the restrictive temperature for additional 2 h (data not shown).

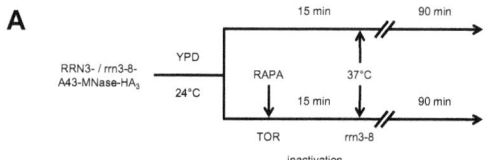

RESULTS

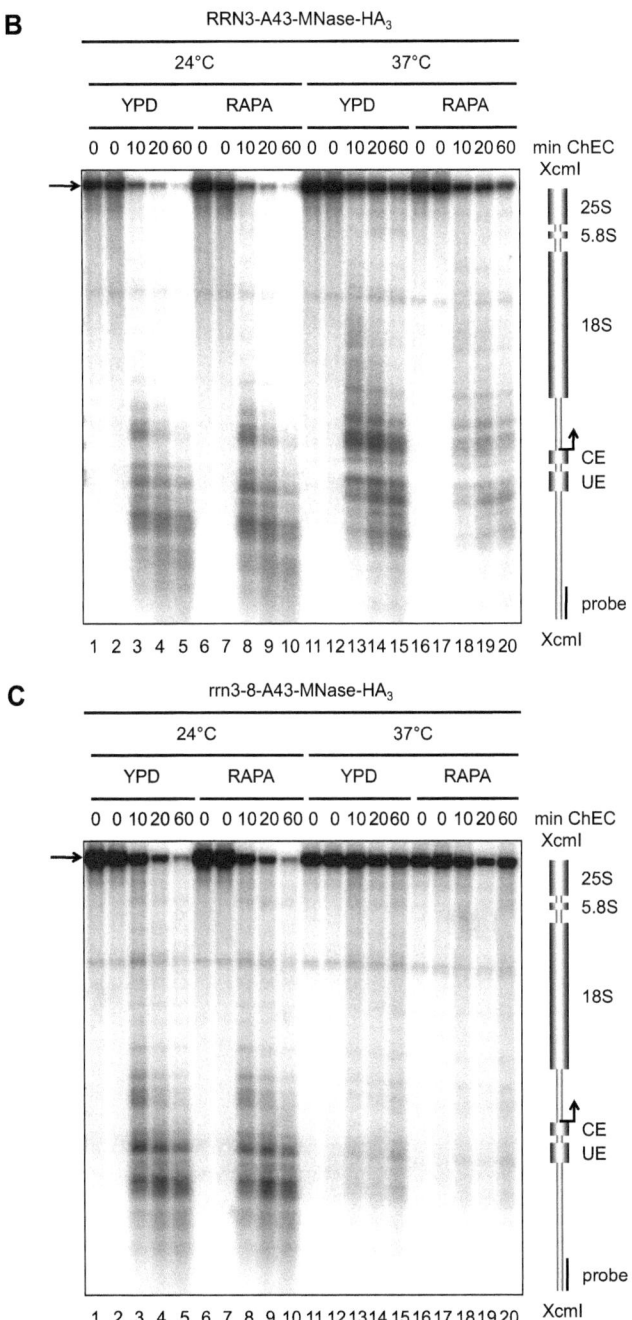

RESULTS

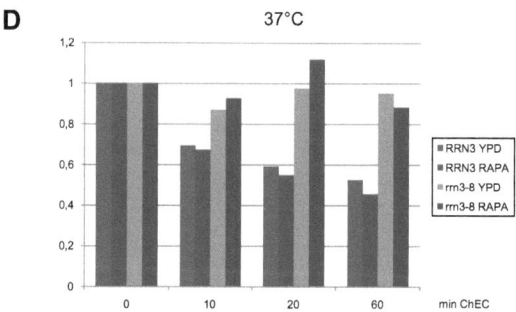

Figure 17. RNA Pol I molecules are not stalled on the rDNA template after 15 min of rapamycin treatment.
(A) Flow chart of the experiment shown in Figure 17B and 17C. Yeast strains Y943 and Y936, both expressing MNase-HA$_3$-tagged Pol I subunit A43 and carrying either an *RRN3* wild type allele or the temperature-sensitive *rrn3-8* allele were grown in YPD at 24°C to mid-log phase. The cultures were split in two parts and rapamycin was added to a final concentration of 200 ng/ml to one half of the culture. After 15 min, samples were withdrawn from the rapamycin treated (RAPA) and untreated (YPD) culture. Cells were crosslinked with 1% formaldehyde, harvested, and analyzed in ChEC experiments. The remainder of the cultures was shifted to 37°C and incubated for another 90 min before cells were crosslinked with 1% formaldehyde, harvested, and analyzed in ChEC experiments.
(B/C) After formaldehyde crosslinking for 15 min at the respective temperature (24°C or 37°C), harvesting of cells, and nuclei preparation, calcium was added to activate ChEC by the MNase fusion proteins. Samples were withdrawn before (0) and at the times indicated on top of the panels. DNA was isolated, linearized with the restriction endonuclease XcmI, separated in an agarose gel, and analyzed in a Southern blot by indirect endlabeling using the rDNA-specific probe XcmI-prom. The autoradiogram of the respective experiment is shown. The sketch on the right shows a map of the corresponding 4.9 kb XcmI rDNA fragment to localize the cleavage events mediated by the MNase fusion proteins. The positions of regulatory elements within the rRNA gene promoter, the upstream element (UE) and the core element (CE), of the 18S, 5.8S and 25S rRNA-coding sequences, of the transcription start site (arrow), and of the target sequence of the radioactive probe are depicted. An arrow on the left marks the full-length XcmI fragment.
(D) Quantitative analysis of the data presented in Figure 17B, lanes 11-20 and Figure 17C, lanes 11-20 was performed with the Multi Gauge software (Fujifilm). In addition to Southern blot hybridization with probe XcmI-prom, the same membrane was also hybridized with probe NUP57 detecting a 4.2 kb fragment encompassing the *RPS23A* gene locus (data not shown). The radioactive signal of the full-length XcmI fragment from this autoradiogram served as an internal loading control, since Pol I is not binding to this Pol II-transcribed gene and therefore unspecific degradation of the full-length transcript could be detected. The radioactive signal of the full-length XcmI fragment of the XcmI-prom-hybridized autoradiogram for each ChEC time point was normalized to the corresponding radioactive signal of the full-length XcmI fragment of the NUP57-hybridized autoradiogram. The extent of degradation is a measure for the association of RNA Pol I with the rRNA gene.

In summary, the results of the ChIP and the ChEC experiments demonstrate that rDNA transcription by RNA polymerase I is only moderately affected after short-term TOR inactivation in yeast cells, whereas pre-rRNA processing is severely and very quickly impaired (Figure 7). Consequently, a process different from Pol I transcription appears to mediate the fast response to TOR inactivation.

3.1.4 Inhibition of translation is sufficient to mimic severe pre-rRNA processing defects observed at early stages of TOR inactivation in yeast cells

Rapamycin is reported to negatively influence not only rDNA transcription but also translation initiation and pre-rRNA maturation in yeast cells (Barbet et al., 1996; Beretta et al., 1996; Berset et al., 1998; Cardenas et al., 1999; Powers and Walter, 1999). Strikingly, inhibition of translation by cycloheximide treatment leads to very similar pre-rRNA processing defects (de Kloet, 1966;

RESULTS

Udem and Warner, 1972; Warner and Udem, 1972). Therefore, it might be possible that impaired translation at early stages of TOR inactivation is the reason for the ribosomal maturation defects. To further elucidate this issue, pulse-chase experiments with [^{3}H]-uracil were performed to directly compare the effects of short-term rapamycin treatment and cycloheximide treatment on rRNA neo-synthesis in yeast cells. To this end, yeast strain RRN3-TAP-A43-HA$_3$ was grown in YPD at 30°C to mid-log phase before the culture was split in three parts and further cultivated in YPD either in the absence or in the presence of rapamycin or cycloheximide, respectively. After 15 min of treatment, same amounts of cells were pulsed for 5 min with [^{3}H]-uracil and chased with an excess of unlabeled uracil for 4 min, 8 min or 16 min. Total RNA was isolated and analyzed by denaturing agarose gel electrophoresis with subsequent Northern blotting and autoradiography (experiment was performed by Alarich Reiter).

Importantly, the cellular [^{3}H]-uracil uptake was affected only moderately and to a similar extent after 15 min of both rapamycin and cycloheximide treatment (data not shown).

As expected, untreated cells showed wild type levels of newly synthesized 35S pre-rRNA, which was subsequently processed to the intermediate 27S and 20S rRNA and finally to the mature 25S and 18S rRNA (Figure 18A, lanes 1-4). Strikingly, in both rapamycin and cycloheximide treated cells, strong maturation defects could be detected resulting in the relative accumulation of labeled 35S pre-rRNA. Additionally, although the initial 35S pre-rRNA-levels were comparable to those observed in untreated cells, the amounts of intermediate rRNA and mature rRNA were strongly reduced in rapamycin treated cells and almost completely lost in cycloheximide treated cells (Figure 18A, lanes 5-12). To illustrate the dimension of the maturation defects in the differently treated yeast cells, the ratio of 35S rRNA to 27S rRNA was calculated and normalized to the ratio in untreated cells (Figure 18B). Furthermore, the incorporation of [^{3}H]-uracil into the mature 25S rRNA was quantified as a measure of ribosome neo-production (Figure 18C). 15 min of TOR inactivation were sufficient to decrease the neo-synthesis of 25S rRNA to 10%, whereas the effect in cells inhibited in translation for the same time was even stronger.

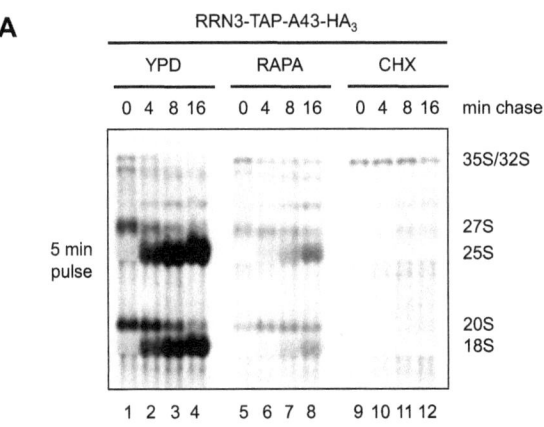

RESULTS

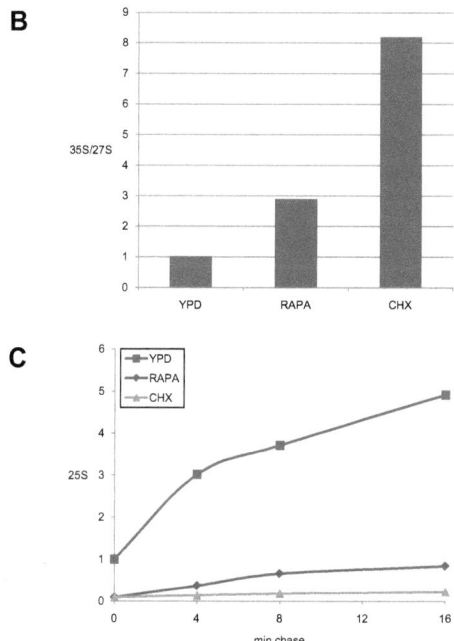

Figure 18. Both short-term rapamycin or cycloheximide treatment lead to severe pre-rRNA processing defects (experiment was performed by Alarich Reiter).
(A) Yeast strain Y658 expressing HA_3-tagged Pol I subunit A43 and TAP-tagged Rrn3p was grown in YPD at 30°C to mid-log phase before the culture was split in three parts and further cultivated in YPD either in the absence or in the presence of rapamycin (200 ng/ml) or cycloheximide (100 μg/ml) for 15 min. Pulse labeling with [^{3}H]-uracil was performed for 5 min followed by a chase with an excess of unlabeled uracil (final concentration 1 mg/ml) for the times indicated above the panel. RNA was isolated, separated in a denaturing agarose gel, and transferred to a positively charged nylon membrane. The autoradiogram shown was obtained after exposure of the membrane treated with EN³HANCE solution. Positions of the different rRNA processing products are indicated on the right.
(B) Quantitative analysis of [^{3}H]-signals presented in Figure 18A was performed prior to EN³HANCE treatment. The ratio of the [^{3}H]-signals of the 35S and 27S pre-rRNAs was calculated after 5 min pulse (0) of each experiment. The 35S/27S rRNA ratio of the cells grown for 15 min in YPD was arbitrarily set to 1.
(C) Incorporation of [^{3}H]-uracil into 25S rRNA was determined by excision of the 25S rRNA bands from an identical blot and analysis by liquid scintillation counting. The values obtained were normalized to the value after 5 min pulse (0) of the culture grown for 15 min in YPD, which was arbitrarily set to 1, and plotted against the time of the chase.

Consistent with previous studies (de Kloet, 1966; Udem and Warner, 1972; Warner and Udem, 1972), these results demonstrate that inhibition of translation by cycloheximide for 15 min leads to severe defects in pre-rRNA maturation and ribosome neo-production comparable to those observed after short-term TOR inactivation by rapamycin.

Additionally, since 15 min of rapamycin treatment caused no alterations in Pol I occupancy at the rDNA locus, Pol I-ChIPs were performed to investigate in detail how the association of RNA Pol I with the rDNA locus is affected upon cycloheximide treatment. The experiment was identical with the one shown in Figure 16C and 16D, except for treatment of the cells with cycloheximide instead of rapamycin.

RESULTS

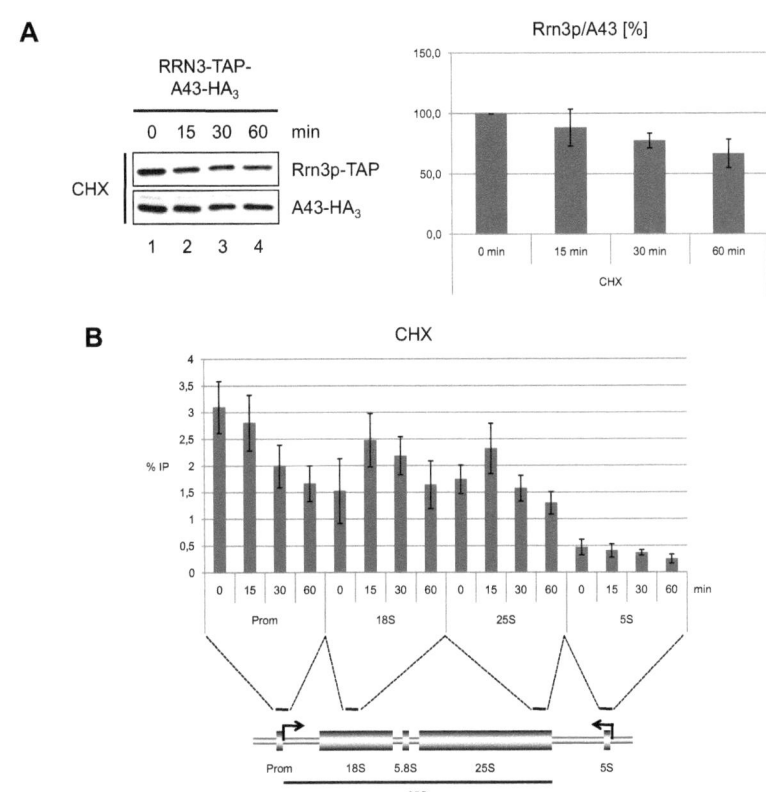

Figure 19. Endogenous Rrn3p-levels and Pol I association with 35S rRNA genes are at most marginally affected after 15 min of cycloheximide treatment.
Yeast strain Y658 expressing HA$_3$-tagged Pol I subunit A43 and TAP-tagged Rrn3p was grown in YPD at 30°C to mid-log phase before cycloheximide was added to a final concentration of 100 μg/ml. At the time points indicated, two samples were withdrawn and cells were either collected and lyzed for subsequent Western blot analysis (A) or treated with 1% formaldehyde at 30°C for 15 min, harvested, lyzed, and sonified for subsequent ChIP experiments (B).
(A) Western blot and quantitative analysis was performed as described in the legend to Figure 16A.
(B) ChIP experiments were performed as described in the legend to Figure 16B.

The results obtained from this time course experiment were strikingly similar to those observed in the corresponding rapamycin experiment (Figure 19). After 15 min of cycloheximide action, just a slight decrease in Rrn3p-levels and in the amount of rRNA gene promoter fragments co-precipitating with Pol I subunit A43 could be detected. Prolonged incubation with the protein synthesis inhibitor for 60 min, however, led to a decline to approximately 60% of both values. The data obtained for A43 crosslinking to the transcribed region of the rDNA were not as easy to be interpreted. They indicated that Pol I recruitment to this region was enhanced after 15 min of cycloheximide treatment compared to the untreated situation. After extended treatment, these levels were diminished again to or slightly below the initial value. Nevertheless, similar to the rapamycin experiment, no significant decrease in RNA Pol I association with the rDNA locus after short-term translation inhibition could be observed, although in both cases the production of mature rRNAs is strongly reduced (Figure 18).

In summary, the above results demonstrate that both short-term TOR inactivation and short-term translation inhibition do not significantly affect the association of Pol I with the rDNA locus, whereas ribosomal subunit production is severely impaired. Accordingly, impaired translation might be sufficient to explain the strong effects of TOR inactivation upon ribosome neo-production and cell growth.

Importantly, impaired protein synthesis predominantly interferes with the abundance of proteins showing both a high expression rate and a high turnover rate. Furthermore, the observation that inhibited translation rapidly leads to a severe pre-rRNA processing defect strongly argues for an abrupt shortage of proteins important for rRNA maturation and ribosome biogenesis. Since the highly expressed ribosomal proteins are characterized by just a small pool size of free, non-ribosome-bound r-proteins due to both a quick assembly rate into pre-ribosomal particles and a high turnover rate, these proteins are very likely candidates (Warner, 1977; Gorenstein and Warner, 1977; Warner et al., 1985; Mitsui et al., 1988; Wittekind et al., 1990). Consistently, conditional shut-down of individual r-proteins is reported to provoke strong defects in rRNA maturation and ribosome biogenesis (Ferreira-Cerca et al., 2005, 2007; Robledo et al., 2008; Pöll et al., 2009).

3.1.5 Short-term TOR inactivation predominantly affects expression of ribosomal proteins whose abundance is important for yeast cell growth

It has been published that within 15 min of rapamycin treatment the mRNA-levels of specifically ribosomal protein genes and ribosome biogenesis genes are significantly and quickly decreased (Powers and Walter, 1999; Jorgensen et al., 2004). In parallel, general translation is shown to be reduced to an extent of 50% (Barbet et al., 1996). Therefore, a sharp decline in the amount of these essential components of ribosome biogenesis under these conditions is very likely.

To investigate whether the neo-production of ribosomal proteins is indeed significantly affected after 15 min of rapamycin treatment, pulse experiments with [^{35}S]-methionine/cysteine were performed. To this end, wild type yeast strain W303-1A was grown in minimal medium depleted of amino acids methionine and cysteine (SCD-met-cys) at 30°C to mid-log phase before the culture was split in six parts and further cultivated in SCD-met-cys either in the absence or in the presence of rapamycin (200 ng/ml) or cycloheximide (0.1 µg/ml; 1 µg/ml; 10 µg/ml; 100 µg/ml). After 15 min of treatment, same amounts of cells were pulsed for 5 min with [^{35}S]-methionine/cysteine. Total protein was extracted and subjected to SDS-PAGE with subsequent coomassie staining and autoradiography. To facilitate the identification of bands corresponding to r-proteins, a sample containing purified 80S ribosomes was loaded on the gel. Strikingly, the pattern derived from the r-proteins of this sample is very similar to that observed in total protein samples, especially in the low-molecular range. To verify the identity of the proteins, the eight

RESULTS

prominent bands migrating with the corresponding mobility of putative ribosomal proteins were excised and analyzed by mass spectrometry. As expected, among 45 proteins identified in these bands, 39 were indeed ribosomal proteins, accounting for approximately 90% of the total protein content.

Strikingly, in the autoradiogram the labeling of these ribosomal proteins is strongly and specifically reduced after 15 min of rapamycin treatment compared to both the expression of overall proteins in this sample and the neo-production of proteins in untreated cells (Figure 20A, compare lane 9 with lane 8). It seems that rapamycin treatment specifically and severely affects neo-production of r-proteins, whereas treatment with increasing cycloheximide concentrations results in a more general decline in the synthesis of all proteins (Figure 20A, compare lane 9 with lanes 10-13). To further document this statement, quantitative analysis of the fast migrating double band (7+8) was performed (Figure 20B). Bands were excised and subjected to liquid scintillation counting. The values obtained were normalized to the coomassie signal of the corresponding band, which served as a loading control. Whereas a significant decline in the neo-synthesis of the respective proteins could be detected in cells treated with 200 ng/ml of rapamycin, just a slight decrease was measureable in cells treated with 0.1 µg/ml of cycloheximide (Figure 20B). Importantly, the overall incorporation of [^{35}S] only marginally differed under these two conditions being about half of the level without drug application (data not shown).

These results correlate well with the results obtained in the pulse-chase experiment shown in Figure 18A. Strong impairment of r-protein expression using 200 ng/ml of rapamycin (Figure 20A, lane 9) led to a strong reduction of pre-rRNA processing (Figure 18A, lanes 5-8). A complete shut-off of r-protein synthesis, however, obtained when treating the cells with 100 µg/ml of cycloheximide (Figure 20A, lane 13) reduced pre-rRNA maturation to background levels (Figure 18A, lanes 9-12).

Recently, a yeast mutant strain was described whose RNA polymerase I molecules remain constitutively competent for the initiation of transcription under stress conditions due to expressing A43-Rrn3p fusion proteins (Laferté et al., 2006). In this CARA mutant, the down-regulation of Pol I transcription upon rapamycin treatment is attenuated, resulting concomitantly in a derepression of Pol II transcription that is restricted to the genes encoding ribosomal proteins. To analyze whether this attenuated decrease in the mRNA-levels of ribosomal proteins leads to increased neo-synthesis of these proteins under these conditions which might diminish the pre-rRNA processing defects, identical pulse experiments were performed including the corresponding wild type strain YPH500 as an appropriate control.

Strikingly, in both cases the same reduction in the expression levels of ribosomal proteins was detected as observed for strain W303-1A (Figure 20C, compare with Figure 20A). Again, this is consistent with results from pulse-chase experiments with [^{3}H]-uracil analyzing the neo-synthesis of rRNA in the strains YPH500 and CARA. The strong impairment of r-protein expression obtained by treating the cells with 200 ng/ml of rapamycin led to a severe pre-rRNA

RESULTS

processing defect in both strains, which could be even aggravated by treating the cells with 100 µg/ml of cycloheximide (data not shown). This led to the conclusion that an attenuated decline in r-protein mRNA-levels is not sufficient to rescue the decrease in the expression levels of ribosomal proteins.

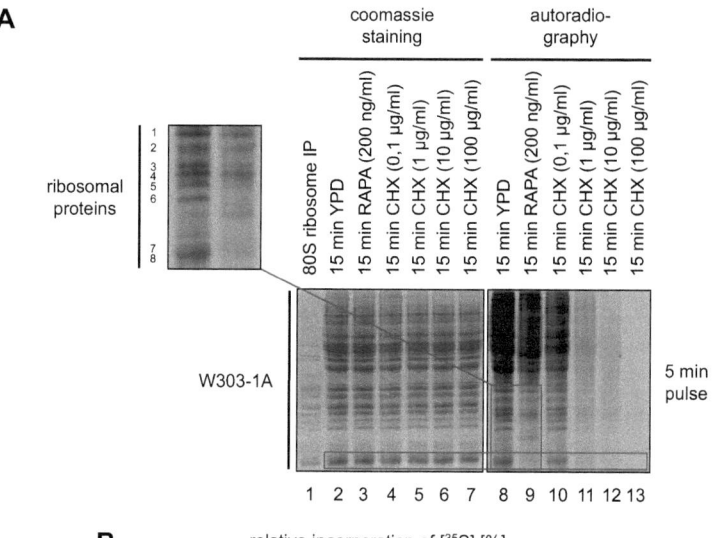

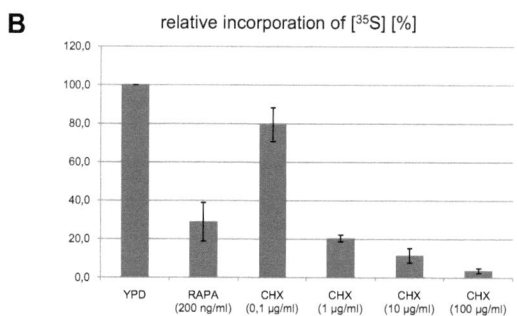

RESULTS

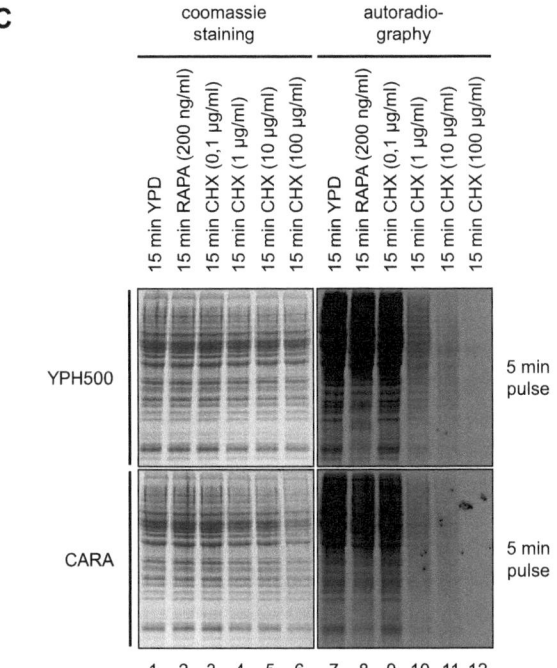

Figure 20. Short-term rapamycin treatment leads to a specific and significant decrease in neo-synthesized r-protein-levels in yeast cells.
(A) Yeast strain Y6 was grown in SCD-met-cys at 30°C to mid-log phase before the culture was split in six parts and further cultivated in SCD-met-cys either in the absence or in the presence of rapamycin (200 ng/ml) or cycloheximide (0.1 µg/ml; 1 µg/ml; 10 µg/ml; 100 µg/ml). After 15 min of treatment, same amounts of cells were pulsed for 5 min with [^{35}S]-methionine/cysteine. Total protein was extracted and subjected to SDS-PAGE with subsequent coomassie staining and autoradiography. A sample containing purified 80S ribosomes was included as a marker for r-proteins. The protein bands which were migrating with the putative mobility of ribosomal proteins and which were therefore analyzed by mass spectrometry are indicated on the left with numbers 1-8.
(B) Incorporation of [^{35}S]-methionine/cysteine into the protein band marked in Figure 20A was determined by excision of the respective band from the coomassie-stained gel and analysis by liquid scintillation counting. The values obtained were normalized to the values of the coomassie-stained bands obtained with the Multi Gauge software (Fujifilm). The level of incorporation of [^{35}S]-methionine/cysteine into the respective protein band of the cells grown for 15 min in YPD was arbitrarily set to 100%. The data represent the mean of at least two independent experiments.
(C) Experiment was essentially performed as described in the legend to Figure 20A, except for using yeast strains Y2172 and Y2171.

The fact that only moderate changes in the level of one single r-protein in the context of gene haploinsufficiency in human disease (Farrar et al., 2008; Lipton and Ellis, 2009, 2010) or in yeast mutants (Abovich et al., 1985; Lucioli et al., 1988; Song et al., 1996; Léger-Silvestre et al., 2005; Deutschbauer et al., 2005) are sufficient to cause pre-rRNA processing and growth defects indicates how drastically a strong reduction of most if not all ribosomal proteins affects yeast cell growth.

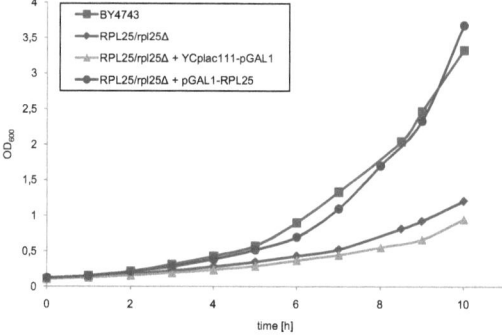

Figure 21. The moderate reduction of the expression level of one single ribosomal protein to approximately 50% results in a significant growth defect.
Diploid yeast strains Y208 and Y990, the latter of which bears a deletion of one *RPL25* allele, and two derivatives of strain Y990 which carry either a control plasmid (YCplac111GAL) or a plasmid expressing the ribosomal protein Rpl25p under the control of a galactose-dependent promoter (pGAL1-RPL25) were grown at 30°C in YPG (Y208) or YPG + geneticin (Y990 + derivatives). Growth was monitored for 10 h by measuring the OD_{600} of the cultures.

To confirm the effect of the reduction of one specific r-protein to presumably 50%, growth of the *RPL25* haploinsufficient diploid yeast strain RPL25/rpl25Δ was monitored. The isogenic wild type strain BY4743 was included in the experiment as an appropriate control. As shown in Figure 21, the growth rate of the mutant strain was significantly lower than that of the corresponding wild type strain. Transforming the mutant strain with a plasmid for galactose-dependent expression of *RPL25* (pGAL1-RPL25) completely rescued growth to wild type rate. However, no significant changes with respect to the growth rate could be observed when this mutant strain is transformed with an identical control plasmid lacking the *RPL25* insert (YCplac111-pGAL1). These results underlined well the importance of maintaining a certain level of this individual ribosomal protein for yeast cell growth.

3.1.6 Nucleolar entrapment of ribosome biogenesis factors in yeast cells is mediated by both rapamycin and cycloheximide treatment as well as by conditional shut-down of ribosomal protein expression

Recently, TOR signaling was reported to be involved in late steps of ribosome maturation, co-transcriptional ribosome assembly and pre-40S ribosome export (Honma et al., 2006; Vanrobays et al., 2008). Rapamycin treatment triggers the nucleolar entrapment of Rrp12p, a cytoplasmic HEAT-repeat/Armadillo-domain and export factor (Oeffinger et al., 2004), and Nog1p, a GTP-binding protein, thereby generating defects in the above processes.

Since the severe impact of TOR inactivation upon ribosome production and cell growth appears to derive from impaired translation and especially from a fast depletion of the free pool of ribosomal proteins, an attempt was made to reproduce the above results and to test in parallel

RESULTS

whether cycloheximide treatment influences the cellular localization of these two ribosome biogenesis factors as well.

Hence, yeast strains RRP12-GFP and NOG1-GFP expressing either GFP-tagged Rrp12p or Nog1p, respectively, were grown in YPD at 30°C to mid-log phase before the cultures were split in three parts and further cultivated either in the absence or in the presence of rapamycin (200 ng/ml) or cycloheximide (100 µg/ml). After 15 min and 30 min, samples were withdrawn, cells were washed, and immediately subjected to live cell imaging. Indeed, in both rapamycin and cycloheximide treated cells a concentration of the GFP-signal to a crescent-shaped region of the nucleus, very likely representing the nucleolus, could be detected, whereas the signal remained dispersed in untreated cells (Figure 22, row 2, 4, 6, and 8, compare central and lower panel with upper panel). The concentration in the nucleolus was more obvious for Rrp12p-GFP, since this factor is distributed all over the cell in normal conditions, whereas Nog1p-GFP is already nuclear in cells cultivated in YPD (Figure 22, row 4 and 8, compare upper and central panel). Furthermore, it appeared that in cells treated with cycloheximide, the subcellular redistribution of the ribosome biogenesis factors occurred even faster than in cells treated with rapamycin (Figure 22, row 2 and 6, compare central and lower panel).

These results suggest that biogenesis factors Rrp12p and Nog1p are no direct targets of the TOR signaling cascade, but their entrapment in the nucleolus is rather a consequence of rapamycin-induced inhibition of protein synthesis, because cycloheximide treatment, which is not reported to affect the TOR pathway, provokes the identical effects.

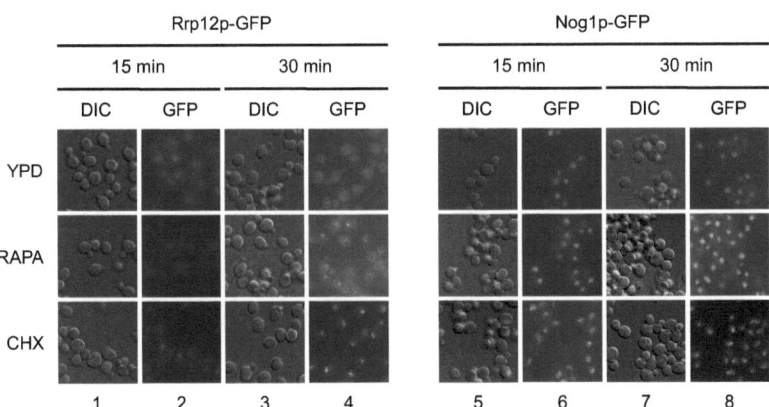

Figure 22. Both rapamycin and cycloheximide treatment result in nucleolar entrapment of ribosome biogenesis factors Rrp12p and Nog1p in yeast cells.
Yeast strains Y1809 and Y1807 expressing either GFP-tagged Rrp12p or Nog1p were grown in YPD at 30°C to mid-log phase before the cultures were split in three parts and further grown in YPD either in the absence or in the presence of rapamycin (200 ng/ml) or cycloheximide (100 µg/ml). At the time points indicated on top of the panels, a sample was withdrawn, washed with SCD, and immediately analyzed by fluorescence microscopy (GFP). Yeast cell morphology was in parallel visualized by differential contrast (DIC).

RESULTS

Next, the hypothesis of rapidly depleted r-proteins generating the observed concentration effects of these ribosome biogenesis factors was tested. Therefore, yeast strains deleted in the chromosomal copies of *RPS5*, *RPS14* or *RPL25* and conditionally expressing the respective wild type allele under the control of the glucose repressible *GAL1* promoter were employed (Ferreira-Cerca et al., 2005; Pöll et al., 2009). Whereas growth of these strains is supported in YPG due to occurring expression of the corresponding ribosomal proteins, a shift of the cells to glucose containing medium inhibits this expression and leads to strong and selective defects in nuclear maturation steps of ribosomal subunit precursors (van Beekvelt et al., 2000, 2001; Ferreira-Cerca et al., 2005, 2007; Pöll et al., 2009). The above strains and the corresponding wild type strains, which served as controls, were genetically modified for constitutive expression of either GFP-tagged Rrp12p or GFP-tagged Nog1p.

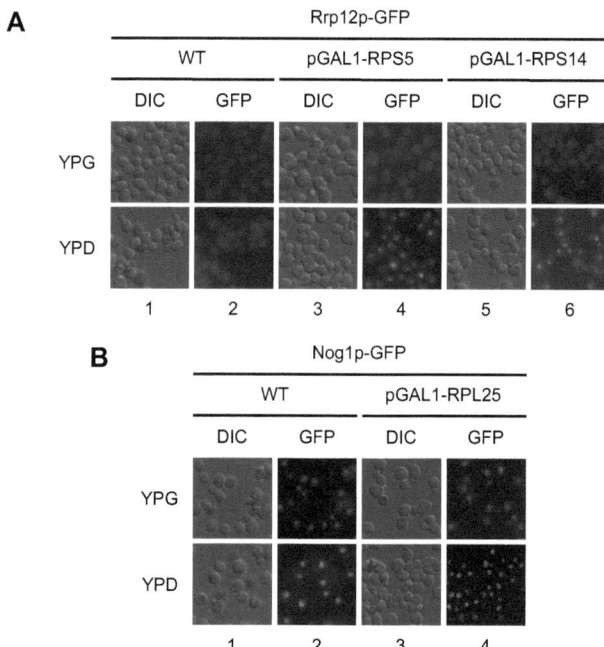

Figure 23. Conditional shut-down of individual ribosomal proteins results in nucleolar entrapment of ribosome biogenesis factors Rrp12p and Nog1p in yeast cells.
(A) Yeast strains Y1809, Y1805, and Y1806 expressing GFP-tagged Rrp12p and expressing ribosomal protein Rps5p or Rps14p either from their respective genomic loci (WT) or from a plasmid under the control of a galactose-dependent promoter (pGAL1-RPS5, pGAL1-RPL25) were grown in YPG at 30°C to mid-log phase before the cultures were split in two parts. Cells were collected, washed, and further cultivated in either YPG or YPD. After 90 min, a sample was withdrawn, washed with SCG or SCD, and immediately analyzed by fluorescence microscopy as described in the legend to Figure 22.
(B) Yeast strains Y1807 and Y1804 both expressing GFP-tagged Nog1p and expressing ribosomal protein Rpl25p either from their respective genomic loci (WT) or from a plasmid under the control of a galactose-dependent promoter (pGAL1-RPL25) were analyzed as described in the legend to Figure 23A.

RESULTS

For the experiment, these yeast strains, RRP12-GFP, NOG1-GFP, pGAL1-RPS5-RRP12-GFP, pGAL1-RPS14-RRP12-GFP, and pGAL1-RPL25-NOG1-GFP, were grown in YPG at 30°C to mid-log phase before the cultures were split in two parts. Cells were collected, washed, and further cultivated either in YPG or YPD. After 90 min, samples were withdrawn, cells were washed, and immediately subjected to live cell imaging. The images obtained showed clearly that the depletion of individual ribosomal proteins in mutant cells grown for 90 min in YPD indeed resulted in nucleolar entrapment of the factors Rrp12p and Nog1p, in contrast to wild type cells (Figure 23A and 23B, row 2, 4 and 6 or row 2 and 4, respectively, compare lower panel with upper panel).

These observations demonstrate that the specific reduction in the abundance of a single ribosomal protein is sufficient to provoke the same effects concerning the subcellular localization of these ribosome biogenesis factors in yeast cells as does rapamycin or cycloheximide treatment.

In summary, the nucleolar entrapment of Rrp12p and Nog1p is obviously not a direct consequence of impaired TOR signaling, but is rather caused indirectly by the rapid and specific depletion of the free pool of ribosomal proteins in this situation which is presumably due to the down-regulation of both transcription of r-protein genes and general translation.

3.2 Effects of overexpression of Rrn3p on RNA polymerase I transcription

3.2.1 *GAL1*-dependent overexpression of Rrn3p results in defects of yeast cell growth

During the preceding studies, it has become obvious that the level of RNA Pol I transcription initiation factor Rrn3p indeed influences transcription of ribosomal RNA genes by RNA polymerase I in yeast. Furthermore, it appeared that enhanced expression of Rrn3p, leading to a larger amount of Pol I-Rrn3p complexes (Figure 12), is accompanied with a decrease in yeast growth rate (Figure 10) (Philippi, 2008). The question arose as to whether and how the overexpression of Rrn3p affects the Pol I system and in turn ribosome production and yeast cell growth or whether the defect in cell growth is mediated by increased Rrn3p-levels in a Pol I-independent manner.

To further elucidate this issue, yeast strain RRN3-Prot.A-A190-HA$_3$ was used expressing Prot.A-tagged Rrn3p and HA$_3$-tagged Pol I subunit A190 both from their genomic loci. The tags are important to allow the analysis of possible alterations in the association of the respective protein with the rDNA locus by chromatin immunoprecipitation experiments.

Transformation of the strain RRN3-Prot.A-A190-HA$_3$ with plasmids pGAL1-RRN3-Prot.A (A/C) or pGAL1-RRN3-Prot.A (2μ) both results in the formation of mutant strains harboring the possibility of a galactose-inducible overexpression of Prot.A-tagged Rrn3p in a co-expression situation. Overexpression from the latter, however, should produce an even higher level of Rrn3p than from the first due to carrying a multi-copy (2μ) and not a single-copy (A/C) plasmid. Another transformation of the above strain with the single-copy plasmid pRRN3-Prot.A (A/C) results in

RESULTS

the formation of a mutant strain constitutively expressing Prot.A-tagged Rrn3p under the control of its endogenous promoter in the same co-expression situation. In the following experiments this strain, hereafter called pRRN3-Prot.A (A/C), was included as a control to be able to monitor the effects of Rrn3p overexpression in the corresponding mutant strains, hereafter called pGAL1-RRN3-Prot.A (A/C) and pGAL1-RRN3-Prot.A (2μ).

The three strains were commonly grown in minimal medium depleted of the amino acid leucine in order to prevent the loss of the plasmids by maintaining selective conditions and containing raffinose as carbon source (SCR-leu) to either permit induction or inhibition of Rrn3p overexpression by adding galactose or glucose to the medium, respectively.

To reproduce the observation that abnormal excess of Rrn3p indeed results in a significant growth defect, pre-cultures of these three strains were grown overnight in SCR-leu at 30°C before different dilutions of the cells were spotted either on SCD-leu (glucose) or SCG-leu (galactose) plates and incubated for 2-4 days at different temperatures (Figure 24).

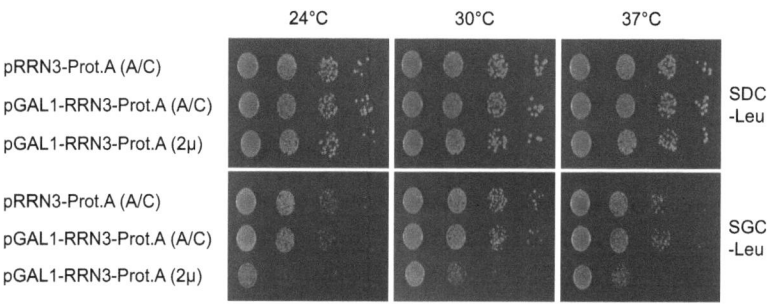

Figure 24. Overexpression of Rrn3p results in defects of yeast cell growth.
Yeast strain Y2127 expressing Prot.A-tagged Rrn3p and HA₃-tagged Pol I subunit A190 both from their genomic loci was transformed either with the single-copy plasmid pRRN3-Prot.A (A/C), with the single-copy plasmid pGAL1-RRN3-Prot.A (A/C), or with the multi-copy plasmid pGAL1-RRN3-Prot.A (2μ), all of them expressing additionally Prot.A-tagged Rrn3p under the control of the endogenous promoter or the GAL1 promoter, respectively. Cells were grown overnight in SCR-leu at 30°C before different dilutions were spotted on SCD-leu or SCG-leu plates containing either glucose or galactose as carbon source and incubated for 3-4 days at the temperatures indicated on top of the panels.

Indeed, growth defects occurred when overexpression of Rrn3p was induced by growth on galactose-containing plates (Figure 24, lower panels) in contrast to plates with the carbon source glucose (Figure 24, upper panels), where the additional expression of Rrn3p is impaired. However, moderate overexpression of this factor in cells harboring the single-copy plasmid pGAL1-RRN3-Prot.A (A/C) was at most slightly affected (Figure 24, lower panel, central spots), whereas strong overexpression in cells with the corresponding multi-copy plasmid pGAL1-RRN3-Prot.A (2μ) caused a significant reduction of the growth rate (Figure 24, lower panel, lower spots). Additionally, it appeared that the effect was more severe when cells were grown at 24°C compared to cells grown at 30°C or 37°C (Figure 24, lower panels, compare left panel with central and right panel). For this reason, in the following experiments cells were grown at 24°C when overexpression of Rrn3p was induced by galactose addition. Additionally, the

RESULTS

overexpression of Rrn3p was induced for at least 8 h, since preliminary experiments showed that after this time the growth defect became detectable by monitoring the optical density of liquid cultures (data not shown).

To get a quantitative estimate, Rrn3p-levels of the three strains were analyzed by Western blotting after cells were grown in SCR-leu+gal for 8 h at 24°C. The cultures were always kept in the logarithmic growth phase (OD_{600} ~ 0.1-0.7) by dilution with SCR-leu+gal.

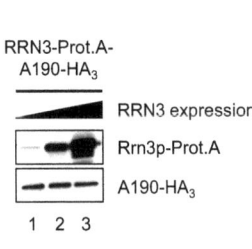
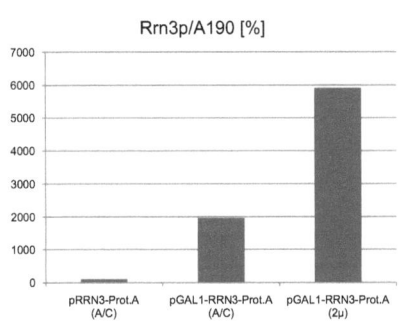

Figure 25. Galactose-dependent expression of Rrn3p in the strain RRN3-Prot.A-A190-HA₃ harboring either plasmid pRRN3-Prot.A (A/C) or plasmids pGAL1-RRN3-Prot.A (A/C) and pGAL1-RRN3-Prot.A (2µ).
Yeast strain Y2127 expressing Prot.A-tagged Rrn3p and HA₃-tagged Pol I subunit A190 and harboring either plasmid pRRN3-Prot.A (A/C), pGAL1-RRN3-Prot.A (A/C), or pGAL1-RRN3-Prot.A (2µ) was grown in SCR-leu at 30°C to mid-log phase before galactose was added to a final concentration of 2% (w/v) and the cultures were shifted to 24°C. The cultures were further grown for 8 h, always keeping the cells in the logarithmic growth phase by dilution with SCR-leu+gal. Cells were collected, lyzed, and same amounts of WCE (9 µg) were analyzed by Western blotting using antibodies directed against the Prot.A-tag of Rrn3p and the HA₃-tag of the Pol I subunit A190, respectively. Quantitative analysis of corresponding signals was performed with the Multi Gauge software (Fujifilm). Rrn3p-Prot.A signals were calculated and normalized to the signals of A190-HA₃, which served as a loading control. The level of Rrn3p of the control strain was arbitrarily set to 100%.

Relative to the level of Pol I subunit A190, the level of Rrn3p could be enhanced 20 fold in the moderate overexpression situation, whereas under strong overexpression conditions even a 60 fold increase could be achieved (Figure 25), which seems to be a critical amount for yeast cell growth (Figure 24, lower panel, lower spots).

3.2.2 Overexpression of Rrn3p leads to increased amounts of Pol I-Rrn3p complexes in yeast cells

As shown before (see section 3.1.2), artificially increased levels of Rrn3p triggers the production of more Pol I-Rrn3p complexes in the cell. To reproduce and to corroborate the results of the preceding experiment, this issue is analyzed here in more detail by handling the examination in a more quantitative way.

To this end, co-immunoprecipitation (CoIP) experiments were performed with the strains pRRN3-Prot.A (A/C), pGAL1-RRN3-Prot.A (A/C), and pGAL1-RRN3-Prot.A (2µ). As described, cells were grown in SCR-leu at 30°C to mid-log phase before galactose was added to the medium and

RESULTS

cells were further cultivated for 8 h at 24°C while keeping them always in the logarithmic growth phase by dilution with SCR-leu+gal. Since only about 2% of all RNA Pol I molecules in a logarithmically growing wild type cell are bound to Rrn3p and therefore competent for transcription initiation (Milkereit and Tschochner, 1998), formaldehyde crosslinking for 15 min was necessary to get co-precipitation of detectable amounts of Rrn3p. Cells were collected, lyzed, and sonified before the Pol I-specific subunit A190 was immunoprecipitated via its HA_3-tag from the chromatin extracts. The associated proteins were eluted with SDS sample buffer and incubated for 30 min at 99°C to reverse the formaldehyde crosslink. Same amounts of the input (3%) and IP fractions (90%) were analyzed by Western blotting. An identical CoIP experiment, except for omitting the required antibodies, was performed and included as an appropriate control.

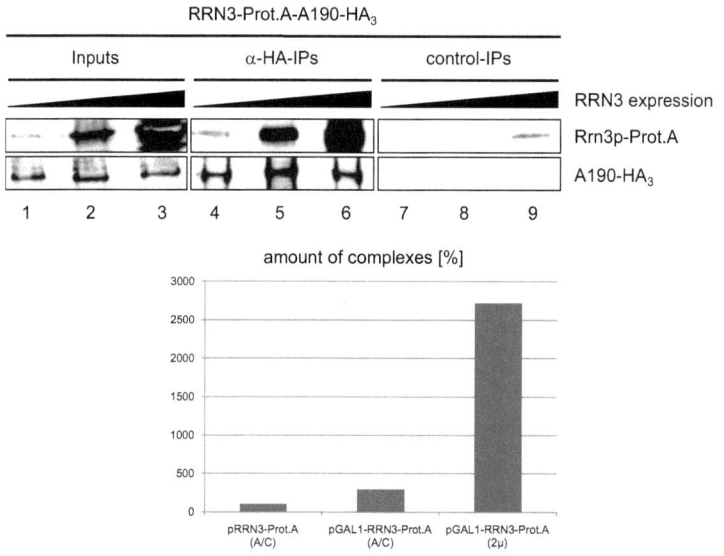

Figure 26. Increased levels of Rrn3p result in elevated levels of Pol I-Rrn3p complexes.
Yeast strain Y2127 expressing Prot.A-tagged Rrn3p and HA_3-tagged Pol I subunit A190 and harboring either plasmid pRRN3-Prot.A (A/C), pGAL1-RRN3-Prot.A (A/C), or pGAL1-RRN3-Prot.A (2µ) was grown in SCR-leu at 30°C to mid-log phase before galactose was added to a final concentration of 2% (w/v) and the cultures were shifted to 24°C. The cultures were further grown for 8 h, always keeping the cells in the logarithmic growth phase by dilution with SCR-leu+gal. Cells were crosslinked with 1% formaldehyde, harvested, lyzed, and sonified. The HA_3-tagged Pol I subunit A190 was immunoprecipitated from the chromatin extracts. Proteins were eluted with SDS sample buffer and the formaldehyde crosslink was reversed by incubation for 30 min at 99°C. 3% of the input and 90% of the IP fraction were analyzed by Western blotting using antibodies directed against the Prot.A-tag of Rrn3p and the HA_3-tag of the Pol I subunit A190, respectively. Quantitative analysis of corresponding signals was performed with the Multi Gauge software (Fujifilm). The ratio of the IP-signals of Rrn3p-Prot.A and A190-HA_3 were calculated and the Rrn3p/A190 ratio of the control strain was arbitrarily set to 100%.

RESULTS

The increasing signal intensities of Rrn3p in the respective input fractions relative to Pol I subunit A190 demonstrate again the occurring overexpression of this factor in the mutant strains upon growth on galactose (Figure 26, left panel).

Regarding the amount of Rrn3p co-precipitating with Pol I subunit A190, it is obvious that increasing Rrn3p-levels are actually accompanied by elevated levels of Pol I-Rrn3p complexes (Figure 26, central panel). By quantification of corresponding signals of the identical samples (data not shown) and subsequent calculation, it could be measured that, by artificially raising the levels of Rrn3p 60 fold, an increase in Pol I-Rrn3p complex formation of more than 25 fold could be generated (Figure 26, quantification).

Importantly, in the control experiment no precipitation of both Pol I and Rrn3p could be detected indicating the high binding specificity of Rrn3p to A190 in the actual experiment (Figure 26, compare right panel with central panel).

3.2.3 ChIP experiments reveal no increase in the association of Pol I with the rDNA locus, but an enhanced level of Rrn3p crosslinking to the rDNA locus when Rrn3p is overexpressed in yeast cells

The next investigation step was to analyze whether the artificially elevated amount of Pol I-Rrn3p complexes by overexpression of Rrn3p leads in turn to an augmented recruitment rate of Pol I and/or Rrn3p to the rRNA genes.

Therefore, ChIP experiments were performed to examine both the crosslinking of Pol I and Rrn3p to the rDNA locus. Yeast strains pRRN3-Prot.A (A/C), pGAL1-RRN3-Prot.A (A/C), and pGAL1-RRN3-Prot.A (2μ) were grown as previously mentioned in SCR-leu at 30°C to mid-log phase before galactose was added to the medium and cells were further cultivated for 8 h at 24°C while keeping them always in the logarithmic growth phase by dilution with SCR-leu+gal. Cells were crosslinked with 1% formaldehyde for 15 min, harvested, lyzed, and sonified. The chromatin extracts were split in two parts and either the HA_3-tagged Pol I subunit A190 or the Prot.A-tagged Rrn3p were immunoprecipitated. After DNA isolation, the relative amounts of specific DNA fragments co-purifying with the respective protein were measured in triplicate real-time PCR reactions using different sets of primer pairs for each of the two proteins. Crosslinking of RNA Pol I was examined to the 35S rRNA gene promoter region and to two regions coding for the 18S rRNA and the 25S rRNA. Crosslinking of Rrn3p was investigated in a similar way to the 35S rDNA promoter region and to one and to two regions coding for the 18S rRNA and the 25S rRNA, respectively. The 5S rRNA-coding region was included in both analyses as an internal control.

RESULTS

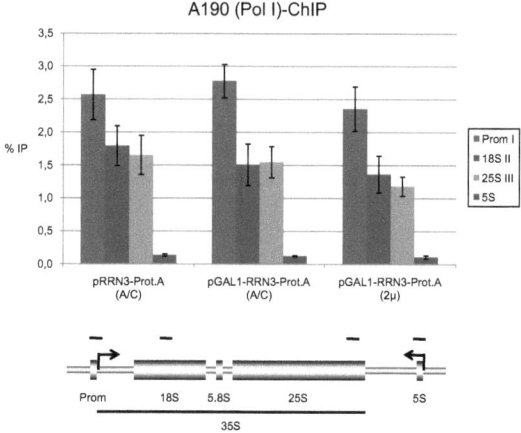

Figure 27. Increased levels of Rrn3p do not lead to an enhanced association of Pol I with the rDNA locus in ChIP experiments.
Yeast strain Y2127 expressing Prot.A-tagged Rrn3p and HA$_3$-tagged Pol I subunit A190 and harboring either plasmid pRRN3-Prot.A (A/C), pGAL1-RRN3-Prot.A (A/C), or pGAL1-RRN3-Prot.A (2μ) was grown in SCR-leu at 30°C to mid-log phase before galactose was added to a final concentration of 2% (w/v) and the cultures were shifted to 24°C. The cultures were further grown for 8 h, always keeping the cells in the logarithmic growth phase by dilution with SCR-leu+gal. Cells were crosslinked with 1% formaldehyde for 15 min, harvested, lyzed, and sonified. The HA$_3$-tagged Pol I subunit A190 was immunoprecipitated from the chromatin extracts. After DNA isolation, the relative amounts of specific DNA fragments co-purifying with the protein were measured in triplicate real-time PCR reactions using primers specific for the rDNA promoter (Prom I), the 18S (18S II) and 25S (25S III) rRNA-coding region as well as for the 5S rRNA gene (5S), which served as an internal control. The data represent the mean of at least three independent ChIP experiments. The positions of the amplified rDNA regions are indicated in a sketch on the bottom of the figure.

The results of the Pol I-ChIP experiment indicated that there was no enhanced recruitment of Pol I, neither to the rDNA promoter region nor to the transcribed region of the rRNA gene, even though more Pol I-Rrn3p complexes were available. Yet, a slight decrease in Pol I association with the rDNA appeared to occur in the situation with strong overexpression of Rrn3p (Figure 27). However, this difference was not drastic and could therefore just be due to measurement inaccuracy. As mentioned before (see section 3.1.2), there are certain limitations of the chromatin immunoprecipitation method in this special case, possibly concealing the existence of increased Pol I occupancy at the rDNA locus. Corresponding ChEC experiments or the Miller chromatin spreading technique might help to overcome this problem.

In contrast to Pol I, a significant increase in the association of Rrn3p with the rRNA gene promoter region but also elevated crosslinking to the transcribed sequence could be detected (Figure 28). However, increased levels of Rrn3p crosslinking to the region coding for the 5S rRNA, which is transcribed by RNA polymerase III, indicated a rise in unspecific binding of the overexpressed protein as well. Nevertheless, when the levels of specific binding were corrected by the respective level of unspecific binding, still a significant increase in the crosslinking of Rrn3p both to the rDNA promoter region and the transcribed sequence could be detected in the course of its overexpression.

RESULTS

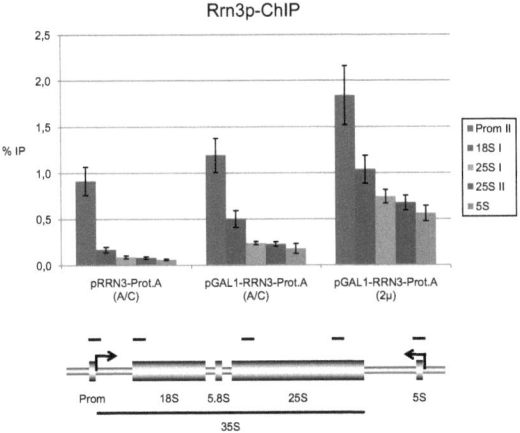

Figure 28. Increased levels of Rrn3p lead to a significantly enhanced association of Rrn3p with the rDNA locus in ChIP experiments.
Yeast strain Y2127 expressing Prot.A-tagged Rrn3p and HA$_3$-tagged Pol I subunit A190 and harboring either plasmid pRRN3-Prot.A (A/C), pGAL1-RRN3-Prot.A (A/C), or pGAL1-RRN3-Prot.A (2µ) was grown as described in Figure 27. Cells were crosslinked with 1% formaldehyde for 15 min, harvested, lyzed, and sonified. The Prot.A-tagged Rrn3p was immunoprecipitated from the chromatin extracts. After DNA isolation, the relative amounts of specific DNA fragments co-purifying with the protein were measured in triplicate real-time PCR reactions using primers specific for the rDNA promoter (Prom II), the 18S (18S I) and 25S (25S I and 25S II) rRNA-coding region as well as for the 5S rRNA gene (5S), which served as an internal control. The data represent the mean of at least three independent ChIP experiments. The positions of the amplified rDNA regions are indicated in a sketch on the bottom of the figure.

Although the possibility of Rrn3p-binding to the rDNA locus independently of Pol I under these overexpression conditions cannot be excluded, Rrn3p was so far described to bind to the promoter region of the rDNA in yeast exclusively when existing in a complex with RNA polymerase I. Therefore, the results obtained from the Rrn3p-ChIP might indicate a higher proportion of Pol I molecules being recruited to the rRNA genes when Rrn3p is overexpressed.

It has been previously demonstrated that Rrn3p dissociates from Pol I when the enyzme switches from initiation to elongation (Schnapp et al., 1993; Brun et al., 1994; Milkereit and Tschochner, 1998; Bier et al., 2004). A large excess of Rrn3p might interfere with this dissociation step leading to problems of Rrn3p-bound Pol I in proper pre-rRNA synthesis and in turn to growth defects as observed in strain pGAL1-RRN3-Prot.A (2µ). A recent study showed that a correlation between a defect in Pol I-Rrn3p complex dissociation and both cell growth and sensitivity to drugs affecting transcription elongation might exist (Beckouet et al., 2008).

RESULTS

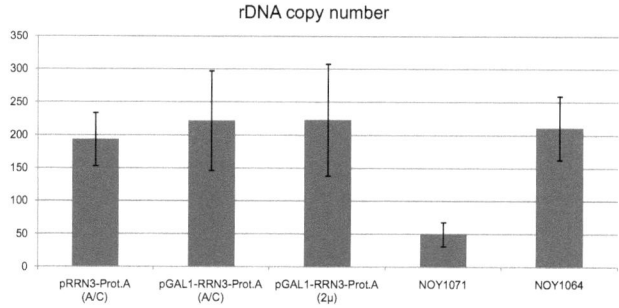

Figure 29. Overexpression of Rrn3p does not lead to a significant alteration in the rDNA copy number.
Yeast strain Y2127 expressing Prot.A-tagged Rrn3p and HA$_3$-tagged Pol I subunit A190 and harboring either plasmid pRRN3-Prot.A (A/C), pGAL1-RRN3-Prot.A (A/C), or pGAL1-RRN3-Prot.A (2µ) and yeast strains Y353 and Y352, carrying approximately 25 or 190 rDNA repeats, respectively, were analyzed as described in Figure 27. After DNA isolation, the content of rDNA was measured in triplicate real-time PCR reactions using primers specific for the 25S rRNA-coding region and normalized to the content of DNA using primers specific for the single copy gene locus *NOC1*.

As described before (see section 3.1.2), artificial variations in the level of Rrn3p did not lead to a significant alteration in the rDNA copy number in yeast cells. This result could be reproduced, since overexpression of Rrn3p showed no substantial impact on the number of rDNA repeats (Figure 29). Yeast strains NOY1071 and NOY1064, carrying approximately 25 or 190 rDNA repeats, respectively, served again as internal controls (Cioci et al., 2003).

3.2.4 Overexpression of Rrn3p does not lead to severe pre-rRNA processing defects or changes in mature rRNA production in yeast cells

Since Rrn3p overexpression might lead to an increased rate of Pol I transcription due to enhanced polymerase loading, the question arose as to whether the presumably thereby generated imbalance between 35S pre-rRNA synthesis and r-protein synthesis leads to pre-rRNA processing defects. Additionally, it is not clear whether there is indeed a problem in Pol I-Rrn3p complex dissociation in this situation which can lead to defects in transcription elongation and pre-rRNA processing as well.

Pulse-chase experiments with [^{3}H]-uracil were performed to directly analyze the effects of Rrn3p overexpression on rRNA neo-synthesis in yeast cells. Therefore, yeast strains pRRN3-Prot.A (A/C), pGAL1-RRN3-Prot.A (A/C), and pGAL1-RRN3-Prot.A (2µ) were grown as described (see section 3.2.3) before same amounts of cells were pulsed for 20 min with [^{3}H]-uracil and chased with an excess of unlabeled uracil for 60 min. Total RNA was isolated and analyzed by denaturing agarose gel electrophoresis with subsequent Northern blotting and autoradiography.

Strikingly, rRNA synthesis seemed not to be critically affected by Rrn3p overexpression. Regardless of the level of Rrn3p, 35S pre-rRNA production and subsequent processing first to the intermediate 27S and 20S rRNA and finally to the mature 25S and 18S rRNA was very comparable in all three strains (Figure 30A). Nevertheless, to reveal potential pre-rRNA processing defects,

RESULTS

the ratio of 35S rRNA to 27S rRNA was calculated and normalized to the corresponding ratio in cells of the control strain. Only slight differences could be observed indicating the absence of severe maturation defects (Figure 30B, compare with Figure 18B). Additionally, the incorporation of [^{3}H]-uracil into the mature 25S rRNA was quantified as a measure of ribosome neo-production. Again, no significant difference between the three strains could be detected (Figure 30C). The neo-synthesis of ribosomes was unaffected by Rrn3p overexpression.

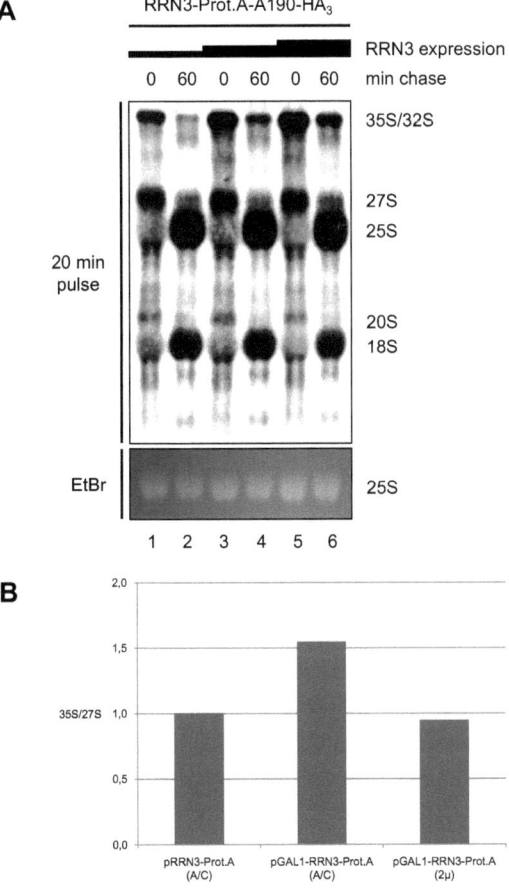

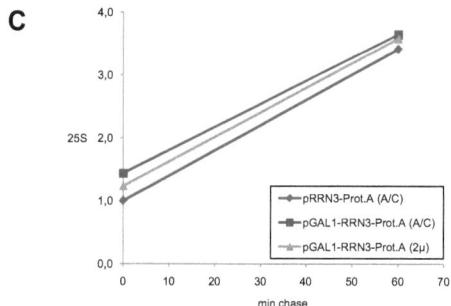

Figure 30. Overexpression of Rrn3p does not result in severe pre-rRNA processing defects or alterations in mature rRNA synthesis.
(A) Yeast strain Y2127 expressing Prot.A-tagged Rrn3p and HA₃-tagged Pol I subunit A190 and harboring either plasmid pRRN3-Prot.A (A/C), pGAL1-RRN3-Prot.A (A/C), or pGAL1-RRN3-Prot.A (2µ) was grown as described in Figure 27. Pulse labeling with [³H]-uracil was performed for 20 min followed by a chase with an excess of unlabeled uracil (final concentration 1 mg/ml) for the times indicated above the panel. RNA was isolated, separated in a denaturing agarose gel, and transferred to a positively charged nylon membrane. The autoradiogram shown was obtained after exposure of the membrane treated with EN³HANCE solution. Positions of the different rRNA processing products are indicated on the right.
(B) Quantitative analysis of [³H]-signals presented in Figure 30A was performed prior to EN³HANCE treatment. The ratio of the [³H]-signals of the 35S and 27S pre-rRNAs was calculated after 20 min pulse (0) of each experiment. The 35S/27S rRNA ratio of the control strain was arbitrarily set to 1.
(C) Incorporation of [³H]-uracil into 25S rRNA was determined by excision of the 25S rRNA bands from an identical blot and analysis by liquid scintillation counting. The values obtained were normalized to the value after 20 min pulse (0) of the control strain, which was arbitrarily set to 1, and plotted against the time of the chase.

If artificially raised levels of Rrn3p indeed generate problems in Pol I-Rrn3p complex dissociation, this will obviously not lead to significant defects in pre-rRNA maturation.

Yet, it seemed that there is a slight increase in 35S pre-rRNA-levels in cells with moderate overexpression of Rrn3p which is even further enhanced under strong overexpression conditions. However, this observation has to be confirmed in additional experiments.

In summary, these results do not explain the growth phenotype of strain pGAL1-RRN3-Prot.A (2µ) cultivated in galactose-containing medium. Besides the probable role of overexpressed Rrn3p in affecting Pol I transcription and ribosome production, definitely other cellular processes could be negatively influenced as well. The question remains open whether the observed growth defect derives indeed from problems in the ribosome synthesis machinery due to an excess of Rrn3p or whether this overexpression impairs different processes in the cell.

3.3 Pol5p, which plays an important role in rRNA synthesis, is a putative interaction partner of Rrn3p

3.3.1 Co-purification of Pol5p in the course of phosphorylation analyses of Rrn3p indicates interaction between the two proteins

As already mentioned, the level of Rrn3p influences Pol I transcription by modulating the formation of Pol I-Rrn3p complexes. However, posttranslational modifications of both Rrn3p and

RESULTS

RNA polymerase I, predominantly phosphorylation and/or dephosphorylation events, are important for Pol I transcription regulation as well. Various studies already revealed the positions and roles of diverse phosphorylation sites of different RNA polymerase I subunits and TIF-IA, the mouse homologue of yeast Rrn3p (Paule et al., 1984; Fath et al., 2001, 2004; Schlosser et al., 2002; Cavanaugh et al., 2002; Yuan et al., 2002; Hirschler-Laszkiewicz et al., 2003; Zhao et al., 2003; Mayer et al., 2004, 2005; Gerber et al., 2008; Hoppe et al., 2009). However, no phosphorylation sites of yeast Rrn3p could be discovered up to now.

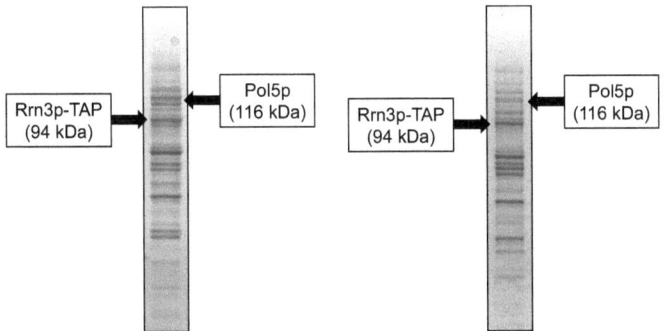

Figure 31. Pol5p is identified in the eluate of two independent multi-step affinity purifications of Rrn3p.
Yeast strain Y2182 expressing TAP-tagged Rrn3p was grown in YPD at 30°C to mid-log phase before cells were collected and lyzed. Rrn3p was affinity precipitated from WCE via its CBP-part of the tag and further purified by anion exchange chromatography using a Mono Q® column before the peak fractions were subjected to SDS-PAGE with subsequent coomassie staining. The bands of the respective proteins identified by mass spectrometry were shown in two independent experiments.

Hence, a further objective of this PhD thesis was the continuation of a project from my diploma thesis in order to reveal such sites of this transcription factor in yeast. To this end, several different large scale affinity purifications of Rrn3p were performed and the fractions containing the enriched protein were mostly subjected to SDS-PAGE to further separate the target protein. The prominent band of Rrn3p and the other dominant protein bands were excised and analyzed by mass spectrometry with respect to either the phosphorylation status in the case of Rrn3p or the protein identity in the case of these putative co-purified interaction partners of Rrn3p. Despite multiple attempts and varying strategies, only the single phosphorylation site of Rrn3p at serine 102 could be confirmed, which I first described in my diploma thesis (see section 4.3) (Steinbauer, 2006). Nevertheless, the protein Pol5p could be identified among the co-purified proteins in two independent multi-step affinity purifications of TAP-tagged Rrn3p during my diploma thesis (Figure 31) (Steinbauer, 2006). Pol5p is shown to play an essential role in the synthesis of ribosomal RNA, presumably by influencing Pol I transcription and/or pre-rRNA processing, since both binding of Pol5p to various sites within the rDNA locus and recruitment of this protein to the 35S pre-rRNA were reported (Shimizu et al., 2002; Yang et al., 2003; Krogan et al., 2004; Nadeem et al., 2006; Wery et al., 2009). Therefore, Pol5p constitutes a *bona fide* interaction partner of Rrn3p. Among the other co-purifying proteins, only the two FACT complex

RESULTS

components Spt16p and Pob3p were identified in one of these purification approaches, likewise exhibiting a direct link to rRNA synthesis due to promoting Pol I transcription elongation (Birch et al., 2009). The question, however, whether this complex indeed interacts with Rrn3p was not investigated further in this thesis.

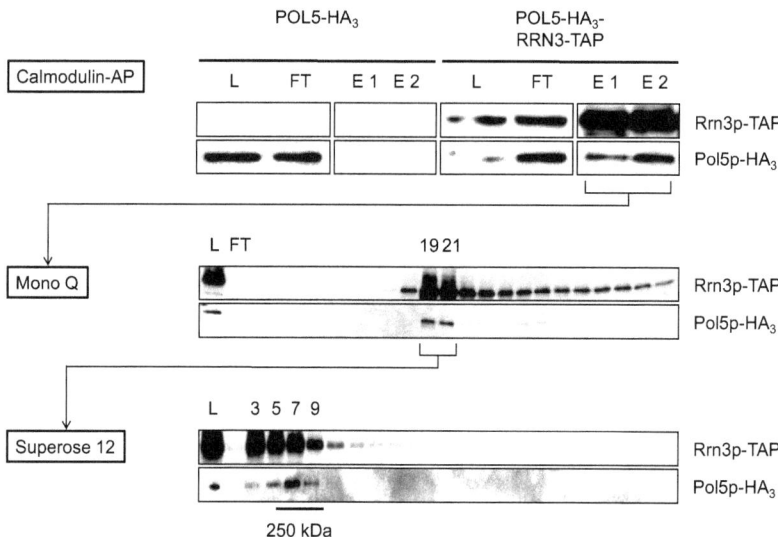

Figure 32. Specific co-purification of Pol5p with Rrn3p within three subsequent purification steps.
Yeast strains Y2179 and Y1798, both expressing HA_3-tagged Pol5p and either untagged or TAP-tagged Rrn3p were grown in YPD at 30°C to mid-log phase before cells were harvested and lyzed. Same amounts of WCE (1.2 g) were incubated with 500 µl calmodulin affinity resin. After washing, proteins bound to the beads were eluted with EGTA. 0.004% of load (L) and flow through (FT) and 0.4% of the eluate (E1/E2) were analyzed by Western blotting using antibodies directed against the HA_3-tag of Pol5p and the Prot.A-tag of Rrn3p, respectively. Pooled eluates were bound to a Mono Q® column and eluted with a linear gradient of 0-1 M sodium chloride. 0.2% of load (L) and flow through (FT) and 3% of every second fraction (19/21) were analyzed by Western blotting using the above antibodies. Pooled peak fractions were separated on a Superose 6® column in a buffer containing 200 mM sodium chloride. 5% of load (L) and 5% of every second fraction (3/5/7/9) were analyzed by Western blotting using the above antibodies.

In my PhD thesis, the putative interaction between Rrn3p and Pol5p should be further investigated. Thus, co-purification of the two proteins was monitored with the adequate yeast strain POL5-HA$_3$-RRN3-TAP. Strain POL5-HA$_3$ was included in the experiment as an appropriate control to look for possible unspecific binding of Pol5p to the calmodulin affinity resin. The strains were grown in YPD at 30°C to mid-log phase before cells were harvested and lyzed. TAP-tagged Rrn3p was affinity precipitated from WCE via the attached calmodulin-binding peptide (CBP) and further purified by anion exchange chromatography and subsequent gel filtration. It was quite obvious that Pol5p co-purified with Rrn3p in all of the three steps of the purification indicating indeed an interaction between the two proteins (Figure 32, upper panel, right hand side, central panel and lower panel). Importantly, no unspecific binding of Pol5p to the calmodulin affinity resin could be detected in the control strain (Figure 32, upper panel, left

RESULTS

hand side). The observation that HA$_3$-tagged Pol5p with a molecular weight of approximately 120 kDa and TAP-tagged Rrn3p with a molecular weight of approximately 90 kDa co-eluted from the gel filtration column in higher molecular weight fractions (~ 250 kDa) than expected for the single proteins provided additional evidence for an existing interaction between the two proteins, probably forming a heterodimer.

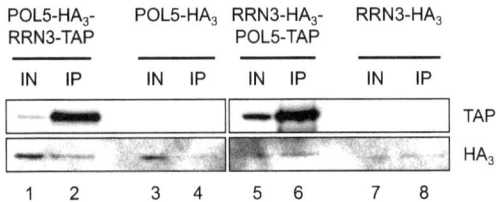

Figure 33. Specific co-precipitation of Pol5p with Rrn3p and *vice versa*.
Yeast strains Y2179 and Y1798 and yeast strains Y653 and Y1799 either both expressing HA$_3$-tagged Pol5p and untagged or TAP-tagged Rrn3p or both expressing HA$_3$-tagged Rrn3p and untagged or TAP-tagged Pol5p were grown in YPD at 30°C to mid-log phase before cells were harvested and lyzed. Rrn3p or Pol5p, respectively, was immunoprecipitated from WCE via its Prot.A-part of the tag. After washing, proteins bound to the beads were eluted with SDS sample buffer. 0.5% of the input (IN) and 50% of the IP (IP) were analyzed by Western blotting using antibodies directed against the HA$_3$-tag or the Prot.A-tag of the respective protein.

Additional co-immunoprecipitation experiments were performed to further confirm the above results. Both, the ability of Pol5p to pull-down Rrn3p and *vice versa* was assayed. To this end, yeast strains POL5-HA$_3$-RRN3-TAP and POL5-HA$_3$ as well as strains RRN3-HA$_3$-POL5-TAP and RRN3-HA$_3$ were grown in YPD at 30°C to mid-log phase before cells were harvested and lyzed. TAP-tagged Rrn3p or Pol5p, respectively, was immunoprecipitated from WCE. The associated proteins were eluted with SDS sample buffer and same amounts of the input (IN) and IP fractions (IP) were analyzed by Western blotting.

The obtained results confirmed an interaction between the corresponding proteins, however, the intensity of the Western blot signals of the HA$_3$-tagged proteins was relatively low (Figure 33, lane 2 and 6). Nevertheless, co-precipitation of Pol5p with Rrn3p and *vice versa* was detectable, whereas in the corresponding control experiments the signals representing unspecific binding were significantly lower than in the actual experiments (Figure 33, lane 4 and 8).

3.3.2 ChIP experiments reveal no association of Pol5p with the rDNA locus

Since Pol5p was reported to bind to the rRNA gene promoter, the 25S rRNA-coding region and the rDNA enhancer region (Shimizu et al., 2002; Nadeem et al., 2006; Wery et al., 2009), additional association of Pol5p to other rDNA regions was assayed by ChIP experiments simultaneously trying to reproduce the published results. Crosslinking of Rrn3p and Pol I subunit A43 to the rDNA locus was explored in parallel, serving as positive controls. Yeast strains POL5-HA$_3$-RRN3-TAP, RRN3-HA$_3$-POL5-TAP, and RRN3-HA$_3$-A43-TAP were grown in YPD at 30°C to mid-log phase before cells were crosslinked with 1% formaldehyde for 15 min, harvested, lyzed, and

RESULTS

sonified. The respective TAP-tagged protein was immunoprecipitated from the chromatin extracts. After DNA isolation, crosslinking of these proteins to the 35S rDNA promoter, to the 18S and the 25S rRNA-coding region and to the enhancer elements EI and EII was examined in triplicate real-time PCR reactions. The 5S rRNA-coding region was included as an internal control at least for the Rrn3p- and Pol I-ChIPs.

Surprisingly, despite investigating Pol5p crosslinking to the regions for which binding was described and even despite using the identical pair of primers for one of these amplicons (25S II), no association of Pol5p could be confirmed with any of the corresponding regions (Figure 34). In summary, no evidence for Pol5p-binding to the rDNA locus was provided in this analysis. Differences in the ChIP protocols of the respective groups could be a possible explanation for the discrepancy of the obtained data. However, Rrn3p crosslinking exclusively to the rRNA gene promoter and Pol I subunit A43-binding to the promoter and the transcribed region was detected as expected (Figure 34).

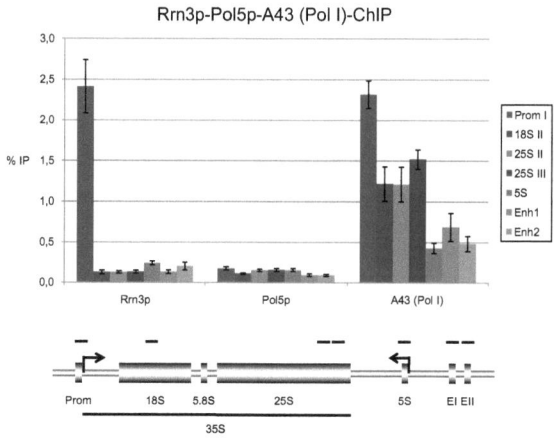

Figure 34. ChIP experiments reveal no association of Pol5p with the rDNA locus.
Yeast strains Y1798, Y1799, and Y2184 expressing either TAP-tagged Rrn3p, Pol5p, or Pol I subunit A43 were grown in YPD at 30°C to mid-log phase before cells were crosslinked with 1% formaldehyde for 15 min, harvested, lyzed, and sonified. Rrn3p, Pol5p, or Pol I subunit A43, respectively, was immunoprecipitated from the chromatin extracts via the Prot.A-part of the tag. After DNA isolation, the relative amounts of specific DNA fragments co-purifying with the protein were measured in triplicate real-time PCR reactions using primers specific for the rDNA promoter (Prom I), the 18S rRNA-coding region (18S II), the 25S rRNA-coding region (25S II and 25S III), the enhancer elements EI and EII (Enh1 and Enh2) as well as for the 5S rRNA gene (5S), which served as an internal control at least for the Rrn3p- and Pol I-ChIPs. The data represent the mean of at least three independent ChIP experiments. The positions of the amplified rDNA regions are indicated in a sketch on the bottom of the figure.

In summary, it is quite obvious that an interaction between at least subpopulations of Pol5p and Rrn3p exists, however, further investigation is required to elucidate the putative binding of Pol5p to the rDNA locus and its role in regulating rRNA synthesis.

4 DISCUSSION

4.1 The role of the proteasome in the down-regulation of Rrn3p-levels upon TOR inactivation

Eukaryotic cells had to evolve several mechanisms to ensure a tight and sensitive regulation of complex processes like ribosome biogenesis. One of these control points is the transcription initiation rate of RNA polymerase I.

In the present PhD thesis, evidence could be obtained suggesting that the yeast ubiquitin-proteasome system is in part involved in the regulation of RNA polymerase I transcription initiation by influencing the abundance of the essential transcription factor Rrn3p (see section 3.1.1).

Ubiquitylation of Rrn3p targets this protein for degradation by the proteasome which contributes to the quick down-regulation of both cellular Rrn3p-levels and Pol I transcription initiation rates following TOR inactivation by rapamycin. Interestingly, Rrn3p appears to be ubiquitylated also in growing cells indicating a constitutive degradation of the protein rather than an induced destruction due to unfavorable growth conditions (Muratani and Tansey, 2003). This hypothesis would implicate a significant reduction in the expression rate of Rrn3p upon TOR inactivation. Indeed, a decrease in *RRN3* mRNA-levels to about 30% of the initial value could be detected within 20 min of rapamycin treatment (Philippi et al., 2010), an observation also described by others (Jorgensen et al., 2004; Huang et al., 2004). Moreover, general translation is reported to decline to roughly 50% of the initial value within 15 min of rapamycin-induced TOR inactivation (Barbet et al., 1996). Consistently, preventing the cellular supply of Rrn3p, either by inhibiting the production of *RRN3* mRNAs in a conditional Pol II mutant strain or by eliminating the translation of *RRN3* mRNAs by cycloheximide, leads to a reduction in the level of Rrn3p with similar kinetics as TOR inactivation. Thus, these observations along with a presumably short half-life of the protein strongly argue for this hypothesis implicating that TOR inactivation affects cellular Rrn3p-levels most probably at the stage of or prior to protein production.

It should be mentioned, however, that the C-terminal Prot.A-tag of Rrn3p could not be entirely excluded to constitute a substrate of the ubiquitylation machinery.

In a different approach, the levels of Rrn3p could be stabilized despite TOR inactivation by shifting cells of the temperature-sensitive proteasome mutant strain *cim3-1* to the restrictive temperature (Philippi et al., 2010). Though, this result contributes to the view of Rrn3p being a genuine target of the ubiquitin-proteasome system, the possible involvement of the tag of Rrn3p in the degradation process could still not be eliminated completely.

Strikingly, TIF-IA, the mouse homologue of Rrn3p, was recently shown to be ubiquitylated as well (Fátyol and Grummt, 2008). This result further corroborates the status of Rrn3p as a direct substrate for ubiquitylation, concomitantly suggesting that ubiquitylation is an evolutionary conserved feature of these proteins. However, contrary to the yeast Rrn3p, the mouse TIF-IA is

DISCUSSION

not degraded upon TOR inactivation but rather redistributed to the cytoplasm (Mayer et al., 2004). This observation indicates that these proteins are differently regulated in their organisms, although fulfilling a very similar function in the cell.

4.2. The role of Rrn3p-levels in the formation of Pol I-Rrn3p complexes upon TOR inactivation

Experiments with a strain whose Rrn3p-levels could artificially be up- or down-regulated showed that there is in fact a correlation between the level of the transcription factor Rrn3p and the amount of initiation-competent Pol I-Rrn3p complexes and their recruitment to the rDNA locus (see section 3.1.2). Similar results could be obtained using strains in which the overexpression of this factor is dependent on galactose (see section 3.2.2).

Several studies showed that the TOR pathway regulates the recruitment of RNA polymerase I to the promoters of active rRNA genes in an Rrn3p-dependent manner (Claypool et al., 2004; Oakes et al., 2006; Philippi et al., 2010). Regardless of whether using the Miller chromatin spreading technique or chromatin immunoprecipitation analysis, the general consensus of this work was that rapamycin-induced TOR inactivation leads to a similar decrease both in Pol I occupancy at the promoter and the transcribed region of the rDNA locus and in Rrn3p occupancy at the rDNA promoter region. In parallel, this down-regulation is accompanied with a decrease in the level of Pol I-Rrn3p complexes (Claypool et al., 2004; Philippi et al., 2010). Thus, this observation suggests that the reduction in the recruitment of Pol I and Rrn3p to the promoter by rapamycin is caused largely, if not entirely, by a decrease in the amount of initiation-competent Pol I-Rrn3p complexes. Concomitantly, a decline in the Rrn3p-levels with similar kinetics could be detected, which is obtained by the combination of a constitutive degradation and a shut-down in the neo-synthesis of the protein (see section 4.1) (Philippi et al., 2010). It is tempting to speculate that the decrease in Pol I-Rrn3p complexes derived exclusively from the respective decrease in Rrn3p-levels. However, although the degradation of Rrn3p contributes definitely to the reduction of Pol I-Rrn3p complexes, there are other mechanisms which play a significant role in this process as well.

Recently, advantage was taken of a strain, referred as to the ΔN strain whose cells express a mutant version of Rrn3p bearing an N-terminal truncation which along with a C-terminal Prot.A-tag confers resistance to rapamycin-induced degradation of the protein. Following TOR inactivation in these cells, the down-regulation of Pol I-Rrn3p complexes, of both Pol I and Rrn3p occupancy at the rDNA locus and of 35S pre-rRNA synthesis could be attenuated, but not completely inhibited (Philippi et al., 2010). Thus, since wild type levels of Rrn3p are maintained in this strain, TOR inactivation must affect complex formation and subsequently Pol I transcription additionally by other means.

The above results indicate that keeping a higher proportion of Pol I-Rrn3p complexes in the cell despite TOR inactivation results in a corresponding higher association of Pol I with the rDNA

DISCUSSION

locus. However, this assumption turned out to be just partially true regarding results obtained from the CARA strain expressing a non-dissociable Pol I-Rrn3p fusion protein (CARA for Constitutive Association of Rrn3 and A43) (Laferté et al., 2006). These cells harbor exclusively Pol I molecules which mimic constitutively active, rapamycin-resistant Pol I-Rrn3p complexes. Similar to the ΔN strain, the CARA strain exhibits elevated levels of Pol I occupancy at the rDNA locus, however, to a larger extent than in the ΔN strain, probably due to preventing both degradation of Rrn3p and its dissociation from RNA polymerase I in this mutant. Yet, a slight but significant decrease in the occupancy of these mutant polymerases at the rDNA locus could still be observed indicating that Pol I-Rrn3p complex formation is not the only mechanism determining the recruitment of the polymerase to the gene. Strikingly, also in CARA cells 35S pre-rRNA synthesis is substantially reduced upon prolonged rapamycin treatment despite maintaining the increased level of Pol I occupancy at the rDNA locus. These results suggest the existence of other mechanisms regulating rRNA production independent from the formation of the Pol I-Rrn3p complex.

4.3. The role of phosphorylation in the formation of Pol I-Rrn3p complexes

Since TOR shows kinase activity, it very likely influences the formation of Pol I-Rrn3p complexes via protein phosphorylation-dephosphorylation cascades. In fact, both in the yeast and the mammalian system, phosphorylation events of Pol I and/or Rrn3p were reported to play essential roles in the regulation of Pol I transcription by affecting the interaction between Pol I and its recruiting factor Rrn3p (Fath et al., 2001; Cavanaugh et al., 2002; Yuan et al., 2002; Hirschler-Laszkiewicz et al., 2003; Zhao et al., 2003; Mayer et al., 2005). Additionally, phosphorylation of TIF-IA was shown to be crucial for mediating the contact between the initiation-competent Pol I complex and the promoter-bound SL1 (Hoppe et al., 2009). Whereas in yeast no reports are published to date describing a TOR-dependent phosphorylation site of either Pol I or Rrn3p, in mammals two serine residues of TIF-IA were found to be posttranslationally modified in a mTOR-dependent manner, thereby affecting complex formation (Mayer et al., 2004).

Interestingly, a recent study in yeast demonstrated that the nuclear localization of Tor1p is critical for 35S pre-rRNA synthesis and that this kinase associates with the promoter region of rRNA genes. Importantly, both the nuclear localization of Tor1p and its association with the rDNA occur in a rapamycin- and starvation-sensitive manner (Li et al., 2006). This observation corroborates the possibility of TORC1-dependent phosphorylation events at the rDNA locus which modulate the transcription initiation rate of RNA polymerase I.

Although diverse phosphorylation sites of the five *in vivo* phosphorylated subunits of yeast Pol I could be identified yet, all of them appear to be non-essential posttranslational modifications (Buhler et al., 1976; Bréant et al., 1983; Gerber et al., 2008). Just the mutation of serine 685 of the largest Pol I subunit A190 to aspartate, which mimics a constitutively phosphorylated state, exhibits synthetic lethality with the deletion of the non-essential Pol I subunit A12.2 (Gerber et

DISCUSSION

al., 2008). The unexpected absence of more severe phenotypes from these mutational analyses could be explained by the possible existence of a cooperation between two or even clusters of phosphorylation sites.

Nevertheless, since Rrn3p was also shown both to be phosphorylated *in vivo* (Fath et al., 2001) and to be a key player in the regulation of Pol I transcription, several phosphorylation analyses were performed in order to identify phosphorylation sites of this protein as it was done previously with RNA polymerase I (Gerber et al., 2008). Despite various attempts and strategies, no additional phosphorylation sites of Rrn3p could be revealed apart from that at serine 102 which I already described in my diploma thesis (Steinbauer, 2006).

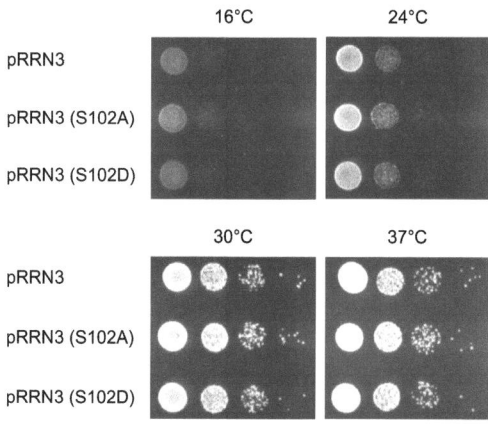

Figure 35. Mutation of phosphorylation site S102 of Rrn3p does not result in defects of yeast cell growth.
Yeast strains Y2176, Y2177, and Y2178 expressing either wild type Rrn3p or mutant Rrn3p whose serine at position 102 is either substituted by alanine (S102A) or aspartate (S102D) were grown overnight in YPD at 30°C before different dilutions were spotted on YPD plates and incubated for 3-4 days at the temperatures indicated on top of the panels.

Identically to the mutational analyses performed for Pol I, the amino acids alanine or aspartate were chosen to substitute serine 102 to mimic either a constitutively unphosphorylated or phosphorylated state, respectively (Gerber et al., 2008). The resulting mutant strains were investigated with respect to their growth ability by performing spot tests on YPD plates at various temperatures in comparison to the control strain. However, regardless of the incubation temperature, the growth behavior of the two mutant strains was indistinguishable from the parent strain (Figure 35) arguing for no essential function of this phosphorylation site.

Despite the above observation, this site, maybe in combination with putative other, yet undefined phosphorylation sites of Rrn3p or with respective sites of RNA polymerase I, still can play a role in the regulation of Pol I transcription.

DISCUSSION

4.4 Uncoupling RNA polymerase I transcription and mature rRNA production after short-term TOR inactivation

As already mentioned, the cellular level of the transcription factor Rrn3p significantly influences RNA polymerase I transcription and thus the rate of rRNA synthesis. A wild type expression level of this factor promotes growth best, whereas both artificially increased and decreased levels of Rrn3p result in distinct growth phenotypes (see sections 3.1.2 and 3.2.1). Importantly, regardless of the level of Rrn3p, even if it is artificially decreased below that of cells treated with rapamycin, all these strains do not cease ribosome production and growth completely. Additional rapamycin treatment of these cells, however, led in all cases to a fast termination of both growth with similar kinetics and mature rRNA production (see section 3.1.2) (Philippi et al., 2010). These results point to another, more drastic and rapid effect of rapamycin- or starvation-induced TOR inactivation on the cells besides down-regulating Pol I transcription via reducing the amount of Rrn3p. Controlling the level of Rrn3p might serve as a tool for the cell to adjust the rate of ribosome synthesis in the long term. At early stages of TOR inactivation, however, a target very likely downstream of Pol I transcription initiation must be affected whose aftermaths dominate and are more severe.

In order to elucidate this issue, the short-term effects of TOR inactivation upon Pol I transcription and rRNA production were investigated. Interestingly, neither the level of Rrn3p nor the association of Pol I with the rDNA locus were substantially decreased after 15 min of rapamycin-induced TOR inactivation (see section 3.1.3). Additionally, the possibility for RNA Pol I of being stalled at the rDNA locus in the course of transcription upon rapamycin treatment could be excluded (see section 3.1.3). At least after 15 min of rapamycin action, polymerase molecules engaged in transcription elongation were still able to quit their started cycle. It should be mentioned, however, that no statements could be made regarding the elongation rate or the processivity of the Pol I molecules following the treatment with rapamycin.

Strikingly, despite the maintenance of 35S pre-rRNA production, there is a drastic decrease in the level of intermediate and mature rRNAs detectable already after 15 min of rapamycin treatment compared to untreated cells (see section 3.1.4). Since TOR inactivation affects not only rRNA gene transcription by Pol I but also very fast translation initiation and pre-rRNA maturation (Barbet et al., 1996; Powers and Walter, 1999), the protein synthesis inhibitor cycloheximide was used to analyze whether a sharp drop in the neo-production of proteins causes the observed pre-rRNA processing defects following TOR inactivation. In fact, 15 min of cycloheximide treatment was sufficient to generate even more severe pre-rRNA maturation defects (see section 3.1.4), a result consistent with diverse earlier reports (de Kloet, 1966; Udem and Warner, 1972; Warner and Udem, 1972). Similar to rapamycin, 15 min of cycloheximide treatment has almost no effect on both cellular Rrn3p-levels and Pol I occupancy at the promoter region of the rDNA locus. It appears that even more Pol I molecules are associated with the transcribed region of the rRNA genes following 15 min of cycloheximide action compared to the untreated situation (see section 3.1.4). A scenario which might account for the observed increase in Pol I

DISCUSSION

crosslinking to the transcribed region could be an improved accessibility of the epitope under these conditions due to enhanced co-transcriptional degradation of nascent transcripts which leads to an elevated precipitation rate in the ChIP experiments. Furthermore, the possibility of an attenuated elongation rate of the Pol I molecules producing similar results could not be excluded.

In summary, TOR-dependent inhibition of protein neo-expression appears to play the dominant role in the drastic down-regulation of ribosome production rather than the decrease in Pol I transcription.

Since ribosomal proteins have to be produced in at least stoichiometric amounts with the ribosomal RNAs in the cell to support proper assembly, processing, and maturation of the ribosomal subunits (Warner, 1999), an impairment in the production of these structural components very likely provokes drastic effects on the neo-synthesis of ribosomes. The occurrence of pre-rRNA processing defect upon depletion of individual r-proteins underline the significance of the above hypothesis (Ferreira-Cerca et al., 2005, 2007; Robledo et al., 2008; Pöll et al., 2009). Further evidence for this statement derives from a yeast strain which exhibits severe growth defects due to an only moderate reduction in the level of a single ribosomal protein in the context of gene haploinsufficiency (see section 3.1.5), an observation already described by others (Abovich et al., 1985; Lucioli et al., 1988; Deutschbauer et al., 2005).

Interestingly, the impact of rapamycin treatment for 15 min is indeed predominantly directed towards the neo-expression of most probably all ribosomal proteins, whereas cycloheximide treatment for the same time affects the overall expression of proteins in the cell (see section 3.1.5). By using limited cycloheximide concentrations, which have a similar impact on the synthesis of ribosomal proteins as rapamycin treatment, a virtually identical defect in the production of mature rRNAs could be obtained (Reiter et al., accepted). Although the CARA strain, when compared to the wild type strain, displays an attenuated decrease in the mRNA-levels specifically of r-proteins upon TOR inactivation (Laferté et al., 2006), their neo-production is equally decreased in both the mutant and the wild type cells after 15 min of rapamycin treatment (see section 3.1.5). Consistent with this finding, in both strains the same drastic defects in pre-rRNA processing and mature rRNA synthesis could be observed under these conditions (data not shown). It should be noted, however, that the decrease in the mRNA-levels of r-proteins in the CARA mutant is only slightly delayed compared to the wild type strain after 15 min of rapamycin treatment. Only after prolonged drug treatment a significant difference between the levels of these mRNAs is established in the two strains.

Further evidence for an early shortage of ribosomal proteins derives from a recent study showing that amino acid starvation leads to a very rapid inhibition of mRNA-splicing in yeast which predominantly affects the expression of r-protein genes (Pleiss et al., 2007). A general depletion of the cellular r-protein-pool was already speculated to link impairments in the nuclear pre-mRNA splicing machinery to ribosome biogenesis defects, since ribosomal protein genes account for almost 50% of the intron-containing genes and about 90% of mRNA splicing is

DISCUSSION

devoted to r-protein transcripts in yeast (Hartwell et al., 1970; Warner and Udem, 1972; Rosbash et al., 1981; Spingola et al., 1999; Warner, 1999).

These observations indicate that the drastic reduction in ribosomal proteins following rapamycin or cycloheximide treatment is sufficient to provoke severe defects in the processing and maturation steps of pre-rRNAs, which are then obviously rapidly degraded.

In this context, the investigation of a recently described *TOR1* mutant strain would be interesting (Li et al., 2006). In this strain, the rapamycin-induced pre-rRNA processing phenotype can be uncoupled from the inhibition of r-protein mRNA synthesis. These observations contradict not necessarily the model of depleted r-proteins generating the immediate shut-off of ribosome production, as it is not clear yet whether or not rapamycin still prevents the translation of the r-protein mRNAs and thus the expression of the ribosomal proteins (Barbet et al., 1996).

Treatment of the yeast cells with rapamycin or cycloheximide not only provokes very related pre-rRNA processing and growth defects, but also the cellular localization of two ribosome biogenesis factors is influenced in a strikingly similar manner (see section 3.1.6). Although the nucleolar entrapment of Rrp12p and Nog1p is described to be caused primarily by the inhibition of the TOR pathway (Honma et al., 2006; Vanrobays et al., 2008), cycloheximide treatment, which affects instantaneously the translation machinery, led to an even accelerated redistribution of these factors to the nucleolus. Considering the depletion of the r-protein-pool to be the reason for the nucleolar entrapment of the two proteins, the observation of this process to occur even faster upon cycloheximide treatment than upon rapamycin treatment is compatible with the differing intensity the two drugs display on the translation machinery. The above speculation of r-protein-levels influencing the localization of these factors could indeed be evidenced, since the conditional depletion of individual r-proteins resulted in the same phenotype as seen for rapamycin and cycloheximide treatment (see section 3.1.6).

4.5 A model for the drastic down-regulation of ribosome production upon TOR inactivation

In summary, the obtained results led to the conclusion that rRNA synthesis and consequently ribosome production in yeast is affected on several levels when TOR signaling is impaired upon rapamycin treatment or starvation.

Firstly, RNA Pol I transcription initiation is negatively influenced. The decrease in the level of the transcription initiation factor Rrn3p leads to a decline in Pol I-Rrn3p complex formation. This in turn results in a corresponding reduction of Pol I transcription. However, the above mechanism represents a rather long-term form of regulation, since only relatively moderate and slow changes in the association of RNA Pol I with the rDNA locus could be detected (Figure 36).

Secondly, RNA Pol I transcription elongation might be affected to a certain extent. The possibility that the Pol I molecules, which are engaged in transcription, are slowed down by TOR inactivation could not be excluded. However, it is clear that these molecules are not stalled on the rDNA template, since they are able to finish their transcription cycle even after approximately 100 min of rapamycin treatment (Figure 36).

Thirdly, and most importantly, pre-rRNA processing is affected very severely and quickly by TOR inactivation. The free pool of ribosomal proteins is very rapidly depleted in cells treated with rapamycin due to a strong and specific down-regulation of r-protein expression. Apparently, this lack causes drastic pre-rRNA processing defects, since ribosomal proteins are indispensable for proper subunit maturation. The incorrectly processed rRNAs are obviously rapidly degraded (Figure 36).

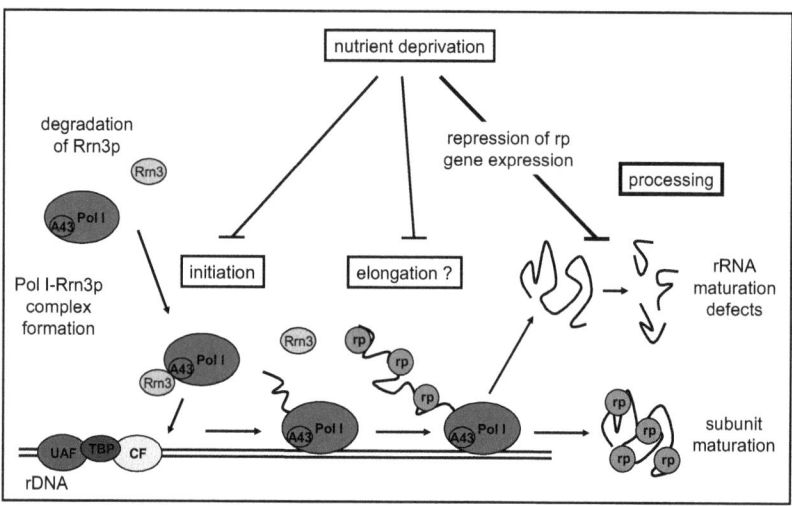

Figure 36. A model for the drastic down-regulation of ribosome production upon TOR inactivation in yeast cells.
Nutrient deprivation affects ribosome production on several levels. Pol I transcription initiation is clearly down-regulated upon TOR inactivation due to a decrease in the level of Rrn3p. Transcription elongation by Pol I might also be affected to some extent, however, the dominant event in the immediate shut-off of ribosome production upon TOR inactivation is the drastic inhibition of pre-rRNA processing mediated by the rapid depletion of ribosomal proteins.

Taken together, these results led to the model that the immediate inhibition of r-protein synthesis is sufficient to explain the drastic pre-rRNA processing defects inducing the acute decline in ribosome neo-production, which in turn is presumably responsible for the observed growth defect in yeast cells upon short-term TOR inactivation.

4.6 Overexpression of Rrn3p and its impact on ribosome biogenesis and yeast cell growth

Since previous observations and the experiments with a strain whose Rrn3p-levels are artificially adjustable indicated that elevated amounts of this factor negatively influence the growth rate of

DISCUSSION

yeast cells, the significance of the overexpression of Rrn3p in the context of Pol I transcription should be further investigated.

Indeed, strong overexpression of this transcription factor led to a substantially decreased growth rate, whereas moderately increased levels of Rrn3p resulted at most in a very slight growth defect if any compared to the control strain (see section 3.2.1). Subsequent analysis revealed that higher levels of Rrn3p generated in fact higher levels of initiation-competent Pol I-Rrn3p complexes (see section 3.2.2), a correlation also true for decreasing amounts of Rrn3p (see section 3.1.2).

The question, however, whether the enlarged pool of Pol I-Rrn3p complexes leads to an augmented recruitment of polymerases to the rDNA locus giving rise to surplus 35S pre-rRNAs could not be answered definitely. Respective ChIP experiments provided ambiguous results. Regardless of the level of Rrn3p, Pol I crosslinking to the promoter and the transcribed region of the rDNA locus was more or less undistinguishable suggesting for no additional polymerases engaged in transcription (see section 3.2.3). Albeit, the corresponding Rrn3p-ChIP indicated that there is indeed an enhanced recruitment of this factor to the rDNA promoter with increasing cellular amounts of Rrn3p (see section 3.2.3). It should be mentioned that unspecific binding of Rrn3p to the 5S rRNA gene is also enhanced. Nevertheless, considering the Rrn3p occupancy at the 5S rDNA region to be background, still a significant increase of specific Rrn3p crosslinking both to the promoter and the transcribed region of the rRNA gene locus could be detected.

Although evidence exist suggesting that hRRN3 is capable of binding to the SL1 and the rRNA gene promoter independently of Pol I in the mammalian system (Miller et al., 2001; Cavanaugh et al., 2002, 2008), corresponding *in vitro* studies showed that Pol I is important for the binding of the CF and thus of Rrn3p to the rDNA promoter in the yeast system (Aprikian et al., 2001). Therefore, the outcome of the Rrn3p-ChIP might argue also for an elevated Pol I occupancy at the rDNA locus. A possible explanation for the failure of detecting more polymerases at the template might be due to certain limitations of the chromatin immunoprecipitation method in this special case. Decreasing occupancy of the enzyme on a distinct rDNA fragment is detectable without limitations, since the amount of the fragment co-precipitating with Pol I gradually declines (see section 3.1.2). However, this is not absolutely true for the opposite case. Both one or five polymerase molecules crosslinked to the same rDNA fragment will just result in the precipitation of this single fragment, thereby masking the assumed fivefold increase in Pol I loading on the rDNA locus (see sections 3.1.2 and 3.2.3).

The appearance of additional rDNA repeats or the activation of further rDNA repeats for transcription would relieve this possible detection problem. However, overexpression of Rrn3p did not result in a significant change in the rDNA copy number (see section 3.2.3). Unfortunately, no clear statements regarding the ratio of actively transcribed to transcriptionally inactive rDNA repeats could be made due to analytical problems, but it seemed that this ratio is not substantially altered as well.

DISCUSSION

Rrn3p is described to dissociate from Pol I during promoter clearance in order to allow efficient transcription elongation by the enzyme (Schnapp et al., 1993; Brun et al., 1994; Milkereit and Tschochner, 1998; Bier et al., 2004). A possible consequence of strongly increased Rrn3p-levels might be an impairment of this dissociation event leading to problems of Pol I in the elongation cycle. This would be consistent with enhanced Rrn3p crosslinking to the transcribed region of the rDNA locus as seen in the Rrn3p-ChIP. Strikingly, a recently published study proposed that there is a connection between the impaired dissociation of the Pol I-Rrn3p complex and problems in both cell growth and transcription elongation (Beckouet et al., 2008).

Interestingly, neither significant pre-rRNA processing defects nor other abnormalities in the production and maturation of ribosomal RNAs were detectable following strong overexpression of Rrn3p which could account for the observed growth phenotype of the respective yeast strain (see section 3.2.4). Only a stepwise increase in the level of 35S pre-rRNA was observed which correlated with rising amounts of Rrn3p. It is difficult to state whether this is the consequence of enhanced Pol I transcription and thus 35S pre-rRNA production or of a minor pre-rRNA processing defect, since 35S pre-rRNA-levels are established by the balanced interplay of pre-rRNA production, pre-rRNA processing, and pre-rRNA degradation.

In any case, further investigation is required, using for instance the Miller chromatin spreading technique or the ChEC method, to clarify the question whether or not Pol I transcription and 35S pre-rRNA production is stimulated by the overexpression of Rrn3p.

It should be noted that the possibility of increased Rrn3p-levels generating defects in processes completely different from ribosome biogenesis cannot be excluded.

4.7 The role of Pol5p in ribosome biogenesis and yeast cell growth

Previous results from my diploma thesis showing twice independently the presence of the factor Pol5p in cellular fractions purified for Rrn3p prompted for a closer investigation of a possible interaction between the two proteins (see section 3.3.1) (Steinbauer, 2006). In fact, at least an interaction between subpopulations of these factors appears to exist, since Pol5p specifically co-purified with Rrn3p within three consecutive purification steps (see section 3.3.1).

Strong evidence exist that Mybbp1a, the corresponding mammalian homologue of Pol5p (Shimizu et al., 2002), interacts likewise with the mammalian RNA Pol I transcription machinery (Hochstatter et al., submitted). However, Mybbp1a was described to bind directly to the Pol I subunit hPAF53/A49. The question whether Pol5p or Mybbp1a additionally interacts with any of the yeast Pol I subunits or the mammalian hRRN3/TIF-IA, respectively, remains open and requires further investigation.

Consistently, an essential role in the synthesis of ribosomal RNA was assigned for both Pol5p and Mybbp1a indicating a conservation of regulatory mechanisms throughout evolution (Shimizu et al., 2002; Nadeem et al., 2006). Both proteins seem to fulfill a dual function in the rDNA metabolism. On the one hand, Pol5p was found to associate with the promoter region, the 25S rRNA-coding region, and the enhancer region of the rDNA locus, thereby presumably

DISCUSSION

influencing Pol I transcription (Shimizu et al., 2002; Nadeem et al., 2006; Wery et al., 2009). On the other hand, this protein could be identified to be a part of a complex required for early pre-rRNA processing events (Krogan et al., 2004). In a related manner, Mybbp1a appears to play both a negative role in Pol I transcription and a positive role in pre-rRNA processing (Hochstatter et al., submitted).

Interestingly, the homologue of Pol5p in the fission yeast *Saccharomyces pombe* appears to interact with a subunit of the cell cycle-regulating complex MBF suggesting a possible role for Pol5p in the coordination of ribosome biogenesis and cell cycle progression (Nadeem et al., 2006).

Mybbp1a, meaning Myb-binding protein 1a, was originally designated due to its ability to bind to the mammalian transcription factor c-Myb (Favier and Gonda, 1994). Being the putative homologue of c-Myb, yeast Reb1p might enable Pol5p to function in Pol I transcription by recruiting the protein to the enhancer region of the rDNA locus where Pol5p crosslinking is actually detected.

Surprisingly, despite investigating the association of Pol5p with the same regions of the rDNA locus where binding was reported, these positive results could not be reproduced (see section 3.3.2) (Shimizu et al., 2002; Nadeem et al., 2006; Wery et al., 2009). Variations in the protocols used for chromatin immunoprecipitation experiments by the different groups provide a possible explanation for this discrepancy.

Although the functional outcome of the depletion of either Pol5p or Mybbp1a is quite similar, being pre-rRNA processing defects and cessation of growth and proliferation (Shimizu et al., 2002; Peng et al., 2003; Hochstatter et al., submitted), further investigation is required to elucidate the conservation of their functions in regulating and/or coordinating ribosome biogenesis, cell growth, and maybe cell cycle progression.

4.8 Outlook

One of the major objectives of this PhD thesis was the investigation of the mechanism(s) regulating the complex formation between RNA polymerase I and Rrn3p. In contrast to the mammalian system, in yeast the parameters involved in the control of this switch essential for RNA polymerase I transcription remain largely uncharacterized. Strikingly, although both Pol I and Rrn3p are present in sufficient amounts in the cell, only a minor fraction of these molecules is assembled into Pol I-Rrn3p complexes.

Alterations in the level of Rrn3p appear to play a distinct role in Pol I-Rrn3p complex formation, however, there are definitely other mechanisms which contribute. Since Rrn3p is a phosphoprotein, changes in the phosphorylation pattern very likely influence its binding affinity to Pol I, a hypothesis already proposed by others (Fath et al., 2001). However, serine 102, the only phosphorylation site of Rrn3p identified so far, turned out to be not essential for yeast cell growth. This is reminiscent to Pol I whose established phosphorylation sites appear likewise to constitute non-essential posttranslational modifications (Gerber et al., 2008).

DISCUSSION

Nevertheless, the identification and investigation of additional phosphorylation sites and/or other posttranslational modifications of both Pol I and Rrn3p is important to further elucidate their role in the regulation of Pol I-Rrn3p complex formation. Crystallization and subsequent X-ray structure analysis of Rrn3p and/or electron microscopy analysis of the Pol I-Rrn3p complex might provide invaluable structural information presumably leading to an improved understanding of Pol I-Rrn3p complex formation.

Finally, the identification of additional interaction partners of Rrn3p and/or Pol I might also contribute to the knowledge of how Pol I-Rrn3p complex formation and thus Pol I transcription is regulated in yeast cells.

5 MATERIAL AND METHODS

5.1 Material

5.1.1 *Saccharomyces cerevisiae* strains

No.	Name	Genotype	Background	Comment	Origin
Y6	W303-1A	ade2-1 can1-100 his3-11,-15 leu2-3,112 trp1-1 ura3-1	W303-1A		(Thomas and Rothstein, 1989)
Y652	cim3-1-RRN3-TAP	ura3-52 leu2-Δ1 his3-Δ200 prc1-1 cim3-1 RRN3::RRN3-TAP (URA3)	YWO365		(Philippi et al., 2010)
Y658	RRN3-TAP-A43-HA$_3$	ura3-52 his3-Δ200 leu2-Δ1 trp1Δ63 lys2-801 ade2-101 prc1-1 RRN3::RRN3-TAP (URA3) RPA43::RPA43-HA$_3$ (HIS3)	YWO365		(Philippi, 2008)
Y2183	RRN3-Prot.A	ade2-1 can1-100 his3-200 leu2-3,112 trp1-1 ura3-1 RRN3::RRN3-Prot.A (HIS3)	BSY420		(Fath et al., 2001)
Y2182	RRN3-TAP	ade2-1 can1-100 his3-200 leu2-3,112 trp1-1 ura3-1 RRN3::RRN3-TAP (TRP1)	BSY420		(Fath, 2002)
Y2189	pRRN3 (URA)	ade2-1 ura3-1 his3-11 trp1-1 leu2-3,112 can1-100 RRN3::HIS3 YCplac33-RRN3 (URA3)	NOY604		(Philippi et al., 2010)
Y667	pTet$_7$-RRN3-Prot.A	ade2-1 ura3-1 his3-11 trp1-1 leu2-3,112 can1-100 RRN3::HIS3 pTet$_7$-RRN3-Prot.A (TRP1)	NOY604		(Philippi et al., 2010)

MATERIAL AND METHODS

Y2180	pNOP1-RRN3-Prot.A	ade2-1 ura3-1 his3-11 trp1-1 leu2-3,112 can1-100 RRN3::HIS3 pNOP1-RRN3-Prot.A (LEU2)	NOY604		(Philippi et al., 2010)
Y2181	pNOP1-RRN3-Prot.A [pGAL1-Myc$_3$-UBI4 [G76A]]	ade2-1 ura3-1 his3-11 trp1-1 leu2-3,112 can1-100 RRN3::HIS3 pNOP1-RRN3-Prot.A (LEU2) pGAL1-Myc$_3$-UBI4 [G76A] (TRP1)	NOY604	obtained by transformation of Y2180 with linearized plasmid pGAL1-Myc$_3$-UBI4 [G76A]	this study
Y2176	pRRN3	ade2-1 ura3-1 his3-11 trp1-1 leu2-3,112 can1-100 RRN3::HIS3 pRS315-RRN3 (LEU2)	NOY604	obtained by transformation of Y2189 with pRS315-RRN3 followed by shuffle (see section 5.2.1.5)	this study
Y2177	pRRN3 (S102A)	ade2-1 ura3-1 his3-11 trp1-1 leu2-3,112 can1-100 RRN3::HIS3 pRS315-RRN3 (S102A) (LEU2)	NOY604	obtained by transformation of Y2189 with pRS315-RRN3 (S102A) followed by shuffle (see section 5.2.1.5)	this study
Y2178	pRRN3 (S102D)	ade2-1 ura3-1 his3-11 trp1-1 leu2-3,112 can1-100 RRN3::HIS3 pRS315-RRN3 (S102D) (LEU2)	NOY604	obtained by transformation of Y2189 with pRS315-RRN3 (S102D) followed by shuffle (see section 5.2.1.5)	this study
Y2113	pRRN3 (URA)-A43-HA$_3$	ade2-1 ura3-1 his3-11 trp1-1 leu2-3,112 can1-100 RRN3::HIS3 YCplac33-RRN3 (URA3) RPA43::RPA43-	NOY604	obtained by transformation of Y2189 with PCR product (2174/2175)	this study

MATERIAL AND METHODS

		HA$_3$ (hphNT1)		from pYM24	
Y2186	pRRN3-Prot.A-A43-HA$_3$	ade2-1 ura3-1 his3-11 trp1-1 leu2-3,112 can1-100 RRN3::HIS3 RPA43::RPA43-HA$_3$ (hphNT1) pRRN3-Prot.A (LEU2)	NOY604	obtained by transformation of Y2113 with pRRN3-Prot.A followed by shuffle (see section 5.2.1.5)	this study
Y2185	pTet$_7$-RRN3-Prot.A-A43-HA$_3$	ade2-1 ura3-1 his3-11 trp1-1 leu2-3,112 can1-100 RRN3::HIS3 RPA43::RPA43-HA$_3$ (hphNT1) pTet$_7$-RRN3-Prot.A (TRP1)	NOY604	obtained by transformation of Y2113 with pTet$_7$-RRN3-Prot.A followed by shuffle (see section 5.2.1.5)	this study
Y943	RRN3-A43-MNase-HA$_3$	ade5 his7-2 leu2-112 trp1-289 ura3-52 RPA43::RPA43-MNase-HA$_3$ (kanMX6)	CG379		(Merz et al., 2008)
Y936	rrn3-8-A43-MNase-HA$_3$	ade5 his7-2 leu2-112 trp1-289 ura3-52 rrn3-8 RPA43::RPA43-MNase-HA$_3$ (kanMX6)	YCC95		(Merz et al., 2008)
Y208	BY4743	his3-1/his3-1 leu2-0/leu2-0 lys2-0/LYS2 met15-0/MET15 ura3-0/ura3-0	BY4743		Euroscarf
Y990	RPL25/rpl25Δ	his3-1/his3-1 leu2-0/leu2-0 lys2-0/LYS2 met15-0/MET15 ura3-0/ura3-0 YOL127w::kanMX4/YOL127w	BY4743		Euroscarf
Y206	BY4741	his3-1 leu2-0 met15-0 ura3-0	BY4741		Euroscarf
Y207	BY4742	his3-1 leu2-0 lys2-0 ura3-0	BY4742		Euroscarf
Y653	RRN3-HA$_3$	his3-1 leu2-0 met15-0 ura3-0 RRN3::RRN3-	BY4741		(Philippi, 2008)

MATERIAL AND METHODS

Y2179	POL5-HA$_3$	HA$_3$ (HIS3) his3-1 leu2-0 met15-0 ura3-0 POL5::POL5-HA$_3$ (kanMX6)	BY4741	obtained by transformation of Y206 with PCR product (1795/1796) from pYM1	this study
Y1798	POL5-HA$_3$-RRN3-TAP	his3-1 leu2-0 met15-0 ura3-0 POL5::POL5-HA$_3$ (kanMX6) RRN3::RRN3-TAP (URA3)	BY4741	obtained by transformation of Y2179 with PCR product (1717/1718) from pBS1539	this study
Y1799	RRN3-HA$_3$-POL5-TAP	his3-1 leu2-0 met15-0 ura3-0 RRN3::RRN3-HA$_3$ (HIS3) POL5::POL5-TAP (URA3)	BY4741	obtained by transformation of Y653 with PCR product (1793/1794) from pBS1539	this study
Y1807	NOG1-GFP	his3-1 leu2-0 met15-0 ura3-0 NOG1::NOG1-GFP (HIS3)	BY4741	obtained by transformation of Y206 with PCR product (2049/2050) from pYM44	this study
Y1809	RRP12-GFP	his3-1 leu2-0 met15-0 ura3-0 RRP12::RRP12-GFP (HIS3)	BY4741	obtained by transformation of Y206 with PCR product (2051/2052) from pYM44	this study
Y323	pGAL1-RPS5	his3-1 leu2-0 ura3-0 lys2-0 YJR123w::kanMX4 pGAL1-RPS5 (LEU2)	BY4742		(Ferreira-Cerca et al., 2005)
Y1805	pGAL1-RPS5-RRP12-GFP	his3-1 leu2-0 ura3-0 lys2-0 YJR123w::kanMX4 pGAL1-RPS5 (LEU2)	BY4742	obtained by transformation of Y323 with PCR product	this study

MATERIAL AND METHODS

		RRP12::RRP12-GFP (hphNT1)		(2051/2052) from pYM25	
Y399	pGAL1-RPS14	his3-1 leu2-0 met15-0 lys2-0 ura3-0 YCR031c::HIS3 YJL191w::kanMX4 pGAL1-RPS14 (LEU2)	BY4741/ BY4742		(Ferreira-Cerca et al., 2005)
Y1806	pGAL1-RPS14-RRP12-GFP	his3-1 leu2-0 met15-0 lys2-0 ura3-0 YCR031c::HIS3 YJL191w::kanMX4 pGAL1-RPS14 (LEU2) RRP12::RRP12-GFP (hphNT1)	BY4741/ BY4742	obtained by transformation of Y399 with PCR product (2051/2052) from pYM25	this study
Y1026	pGAL1-RPL25	his3-1 leu2-0 ura3-0 YOL127w::kanMX4 pGAL1-RPL25 (LEU2)	BY4741/ BY4742		(Pöll et al., 2009)
Y1804	pGAL1-RPL25-NOG1-GFP	his3-1 leu2-0 ura3-0 YOL127w::kanMX4 pGAL1-RPL25 (LEU2) NOG1::NOG1-GFP (hphNT1)	BY4741/ BY4742	obtained by transformation of Y1026 with PCR product (2049/2050) from pYM25	this study
Y624	A190-HA$_3$	ade2-1 ura3-1 leu2-3,112 trp1-1 his3-11 can1-100 RPA190::RPA190-MNase-HA$_3$ (kanMX6)	NOY505		(Merz et al., 2008)
Y2127	RRN3-Prot.A-A190-HA$_3$	ade2-1 ura3-1 leu2-3,112 trp1-1 his3-11 can1-100 RPA190::RPA190-MNase-HA$_3$ (kanMX6) RRN3::RRN3-Prot.A (HIS3)	NOY505	obtained by transformation of Y624 with PCR product (242/243) from pYM10	this study
Y2184	RRN3-HA$_3$-A43-TAP	ade2-101 ura3-52 lys2-801 trp1-63 his3-200 leu2-1 RPA43::RPA43-TAP	YPH499		Philippi, Anja

MATERIAL AND METHODS

		(TRP1) RRN3::RRN3-HA$_3$ (HIS3)			
Y2172	YPH500	ade2-101 his3-200 leu2-1 lys2-801 trp1-63 ura3-52	YPH500		(Sikorski and Hieter, 1989)
Y2171	CARA	ade2-101 his3-200 leu2-1 lys2-801 trp1-63 ura3-52 RRN3::HIS5 RPA43::kan pGEN-RRN3-A43 (TRP1)	YPH500		(Laferté et al., 2006)
Y353	NOY1071	ade2-1 ura3-1 his3-11 trp1-1 leu2-3,112 can1-100 FOB1::HIS3	NOY1071		(Cioci et al., 2003)
Y352	NOY1064	ade2-1 ura3-1 his3-11 trp1-1 leu2-3,112 can1-100 FOB1::HIS3	NOY1064		(Cioci et al., 2003)

5.1.2 *Escherichia coli* strains

Name	Genotype	Origin
XL1-Blue	endA1 gyrA96(nal^R) thi-1 recA1 relA1 lac glnV44 F'[::Tn10 proAB$^+$ lacI^q Δ(lacZ)M15] hsdR17(r$_K^-$ m$_K^+$)	Stratagene
BL21(DE3)	F$^-$ ompT gal dcm lon hsdS$_B$(r$_B^-$ m$_B^-$) λ(DE3 [lacI lacUV5-T7 gene 1 ind1 sam7 nin5])	Stratagene

5.1.3 Plasmids

Plasmid	Gene	Description	Derived from	Origin
YCplac33-RRN3	RRN3	URA3, Amp^r, ORI, ARS1, CEN4		(Philippi et al., 2010)
pRRN3-Prot.A (A/C)	RRN3	LEU2, Amp^r, ORI, ARSH4, CEN6	pNOP1-RRN3-Prot.A (see section 5.2.3.12.1)	this study
pTet$_7$-RRN3-Prot.A	RRN3	TRP1, Amp^r, ORI, ARS1, CEN4		(Philippi et al., 2010)
pGAL1-RRN3 (A/C)	RRN3	LEU2, Amp^r, ORI, ARSH4, CEN6		Seufert, Wolfgang
pGAL1-RRN3-	RRN3	LEU2, Amp^r, ORI,	pGAL1-RRN3	this study

MATERIAL AND METHODS

Prot.A (A/C)		ARSH4, CEN6	(A/C) (see section 5.2.3.12.2)	
pGAL1-RRN3 (2μ)	RRN3	LEU2, Amp^r, ORI, 2μ		Seufert, Wolfgang
pGAL1-RRN3-Prot.A (2μ)	RRN3	LEU2, Amp^r, ORI, 2μ	pGAL1-RRN3 (2μ) (see section 5.2.3.12.2)	this study
pNOP1-RRN3-Prot.A	RRN3	LEU2, Amp^r, ORI, ARSH4, CEN6		(Philippi et al., 2010)
pGAL1-Myc$_3$-UBI4 [G76A]	UBI4	TRP1, Amp^r, ORI		Seufert, Wolfgang
pRS315-RRN3	RRN3	LEU2, Amp^r, ORI, ARSH4, CEN6		Milkereit, Philipp
pRS315-RRN3 (S102A)	RRN3	LEU2, Amp^r, ORI, ARSH4, CEN6	pRS315-RRN3 (see section 5.2.3.12.3)	this study
pRS315-RRN3 (S102D)	RRN3	LEU2, Amp^r, ORI, ARSH4, CEN6	pRS315-RRN3 (see section 5.2.3.12.3)	this study
pGEN-RRN3-A43	RRN3/RPA43	TRP1, Amp^r, ORI, 2μ		(Laferté et al., 2006)
YCplac111-pGAL1		LEU2, Amp^r, ORI, ARS1, CEN4		(Ferreira-Cerca et al., 2005)
pGAL1-RPS5	RPS5	LEU2, Amp^r, ORI, ARS1, CEN4		(Ferreira-Cerca et al., 2005)
pGAL1-RPS14	RPS14	LEU2, Amp^r, ORI, ARS1, CEN4		(Ferreira-Cerca et al., 2005)
pGAL1-RPL25	RPS25	LEU2, Amp^r, ORI, ARS1, CEN4		(Pöll et al., 2009)
pYM1	-	kanMX6, Amp^r, ORI		(Knop et al., 1999)
pYM10	-	HIS3, Amp^r, ORI		(Knop et al., 1999)
pYM24	-	hphNT1, Amp^r, ORI		(Janke et al., 2004)
pYM25	-	hphNT1, Amp^r, ORI		(Janke et al., 2004)
pYM44	-	HIS3, Amp^r, ORI		(Janke et al., 2004)
pBS1539	-	URA3, Amp^r, ORI		(Puig et al., 2001)

MATERIAL AND METHODS

5.1.4 Oligonucleotides

No.	Name	Sequence	Gene	Function
242	Rrn3-pYM-for	GGAGTGAAGCAAGCGGGGAATATGAAAGTGATGGGTCGGATGACCGTACGCTGCAGGTCGAC	RRN3	primer to obtain amplicon of pYM10 for genomic integration of Prot.A (HIS3)
243	Rrn3-pYM-rev	TAGTTTGTGACGGGCATGTCTCGAAGATACCTATGAAAAAGACCATCGATGAATTCGAGCTCG	RRN3	primer to obtain amplicon of pYM10 for genomic integration of Prot.A (HIS3)
710	M1	TGGAGCAAAGAAATCACCGC	rDNA	primer used for qPCR amplifying a region in 25S rDNA (25S III)
711	M2	CCGCTGGATTATGGCTGAAC	rDNA	primer used for qPCR amplifying a region in 25S rDNA (25S III)
712	M3	GAGTCCTTGTGGCTCTTGGC	rDNA	primer used for qPCR amplifying a region in 18S rDNA (18S II)
713	M4	AATACTGATGCCCCCGACC	rDNA	primer used for qPCR amplifying a region in 18S rDNA (18S II)
920	5SChIPF1	GCCATATCTACCAGAAAGCACC	rDNA	primer used for qPCR amplifying a region in 5S rDNA (5S)
921	5SChIPR1	GATTGCAGCACCTGAGTTTCG	rDNA	primer used for qPCR amplifying a region in 5S rDNA (5S)
969	PromChIPF2	TCATGGAGTACAAGTGTGAGGA	rDNA	primer used for qPCR amplifying rDNA promoter region (Prom I)
970	PromChIPR1	TAACGAACGACAAGCCTACTC	rDNA	primer used for qPCR amplifying rDNA promoter region (Prom I)
1717	fp Rrn3-TAP-URA3	GGAATACTACATAGAGTGGAGTGAAGCAAGCGGGGAATATGAAAGTGATGGGTCGGATGACTCCATGGAAAAGAGAAG	RRN3	primer to obtain amplicon of pBS1539 for genomic integration of TAP (URA3)
1718	rp Rrn3-TAP-URA3	GACTACAGTGTATCATTAGTTTGTGACGGGCATGTCTCGAAGATACCTATGAAAAAGACCTACGACTCACTATAGGG	RRN3	primer to obtain amplicon of pBS1539 for genomic integration of TAP (URA3)
1793	fp Pol5-TAP-URA3	TTCATCAATTGGCTATCTTCAAAAAAGCAAACTGTAATGGATAAGGAATCCATGGAAAAGAGAAG	POL5	primer to obtain amplicon of pBS1539 for genomic integration of TAP (URA3)

1794	rp Pol5-TAP-URA3	GATCAACATACATAATCCGATTTTGAGGAGTGATAAAATATACTTAGATACGACTCACTATAGGG	POL5	primer to obtain amplicon of pBS1539 for genomic integration of TAP (URA3)
1795	fp Pol5-HA$_3$-kanMX6	TCATCAATTGGCTATCTTCAAAAAAGCAAACTGTAATGGATAAGGAACGTACGCTGCAGGTCGAC	POL5	primer to obtain amplicon of pYM1 for genomic integration of HA$_3$ (kanMX6)
1796	rp Pol5-HA$_3$-kanMX6	TCAACATACATAATCCGATTTTGAGGAGTGATAAAATATACTTAGAATCGATGAATTCGAGCTCG	POL5	primer to obtain amplicon of pYM1 for genomic integration of HA$_3$ (kanMX6)
2049	Nog1-GFP for S3	TTATTCAGTGGTAAGCGTGGTGTCGGTAAGACAGATTTCCGTCGTACGCTGCAGGTCGAC	NOG1	primer to obtain amplicon of pYM25/pYM44 for genomic integration of GFP (HIS3)
2050	Nog1-GFP rev S2	TTTTAAAGTACATGCAAAAGAAAGAATAAGAAGTAGAGAAAATCGATGAATTCGAGCTCG	NOG1	primer to obtain amplicon of pYM25/pYM44 for genomic integration of GFP (HIS3)
2051	Rrp12-GFP for S3	CATAATAAGAAAGGTCCAAAGTTCAAATCTAGAAAAAAATTACGTACGCTGCAGGTCGAC	RRP12	primer to obtain amplicon of pYM25/pYM44 for genomic integration of GFP (HIS3)
2052	Rrp12-GFP rev S2	CTCCAGGTGTGTAATTAGCCATATTGCTCAGTTTCAATCTTATCGATGAATTCGAGCTCG	RRP12	primer to obtain amplicon of pYM25/pYM44 for genomic integration of GFP (HIS3)
2137	RRN3-SOE R1A	CTATCCTGTTGATATTAGCAGATAAGATATCCAAA	RRN3	primer to obtain phospho-mutant allele of RRN3 (see section 5.2.3.12.3)
2138	RRN3-SOE R1D	CTATCCTGTTGATATTATCAGATAAGATATCCAAA	RRN3	primer to obtain phospho-mutant allele of RRN3 (see section 5.2.3.12.3)
2139	RRN3-SOE F2A	TTTGGATATCTTATCTGCTAATATCAACAGGATAG	RRN3	primer to obtain phospho-mutant allele of RRN3 (see section 5.2.3.12.3)
2140	RRN3-SOE F2D	TTTGGATATCTTATCTGATAATATCAACAGGATAG	RRN3	primer to obtain phospho-mutant allele of RRN3 (see section 5.2.3.12.3)

2141	RRN3-SOE F1	TCGAGGTCGACGGTATCGATAAGCT	RRN3	primer to obtain phospho-mutant allele of RRN3 (see section 5.2.3.12.3)
2142	RRN3-SOE R2	ATATTCCCCGCTTGCTTCACTCCAC	RRN3	primer to obtain phospho-mutant allele of RRN3 (see section 5.2.3.12.3)
2174	fp A43-HA$_3$-kanMX6	GAGATCGTATACGAGGAAACACCAGTGAAAGCAATGATGGTGAATCGAGTGATAGTGATCGTACGCTGCAGGTCGAC	RPA43	primer to obtain amplicon of pYM24 for genomic integration of HA$_3$ (kanMX6)
2175	rp A43-HA$_3$-kanMX6	TTATCCTATATCAATAACGTATATCTTTATTTGTTTTGATTTTTTCTCATTTTTCCCGTCATCGATGAATTCGAGCTCG	RPA43	primer to obtain amplicon of pYM24 for genomic integration of HA$_3$ (kanMX6)
2234	RRN3 fo SacI	CTCACTATAGGGCGAATTGGAGCTC	RRN3	primer to obtain amplicon to clone RRN3 promoter
2235	RRN3 re NdeI	CCGCTTGCTTCACTCCACTCTATGT	RRN3	primer to obtain amplicon to clone RRN3 promoter
2472	NdeI fo	TCTTTAGCCACTCGACAACA	RRN3	primer to obtain amplicon to clone Prot.A-tagged RRN3
2473	HindIII re	TACGCCAAGCTCGGAATTAA	RRN3	primer to obtain amplicon to clone Prot.A-tagged RRN3
2477	rDNA3 Th fo	GTGTGAGGAAAAGTAGTTGGGAGGTA	rDNA	primer used for qPCR amplifying rDNA promoter region (Prom II)
2478	rDNA3 Th re	GACGAGGCCATTTACAAAAACATAAC	rDNA	primer used for qPCR amplifying rDNA promoter region (Prom II)
2479	rDNA4 Th fo	CTTGTCTCAAAGATTAAGCCATGC	rDNA	primer used for qPCR amplifying a region in 18S rDNA (18S I)
2480	rDNA4 Th re	ACCACAGTTATACCATGTAGTAAAGGAACT	rDNA	primer used for qPCR amplifying a region in 18S rDNA (18S I)
2481	rDNA8 Th fo	GGTGGTAAATTCCATCTAAAGCTAAATATT	rDNA	primer used for qPCR amplifying a region in 25S rDNA (25S I)
2482	rDNA8 Th re	CACGTACTTTTTCACTCTCTTTTCAAA	rDNA	primer used for qPCR amplifying a region in 25S rDNA (25S I)
2483	rDNA11 Th fo	AAAGAAGACCCTGTTGAGCTTGA	rDNA	primer used for qPCR amplifying a region in 25S rDNA (25S II)
2484	rDNA11 Th re	GTATTTCACTGGCGCCGAA	rDNA	primer used for qPCR amplifying a region in 25S rDNA (25S II)

MATERIAL AND METHODS

817	rDNA-709_for	GAGGGACGGTTGAAAGTG	rDNA	primer to obtain template for Southern probe preparation from yeast genomic DNA
818	rDNA-461_re	ATACGCTTCAGAGACCCTAA	rDNA	primer to obtain template for Southern probe preparation from yeast genomic DNA
1167	Nup57+759-for	TGGATCTAATTTACAGCAGCA	NUP57	primer to obtain template for Southern probe preparation from yeast genomic DNA
1168	Nup57+1012-rev	CCTGATCCCACTCTTCTTGA	NUP57	primer to obtain template for Southern probe preparation from yeast genomic DNA

5.1.5 Probes

Name	Synthesis	Gene	Restriction Enzyme	Fragment Size
XcmI-prom	PCR from yeast genomic DNA using primer 817/818	rDNA	XcmI	4.9 kb
NUP57	PCR from yeast genomic DNA using primer 1167/1168	RPS23A	XcmI	4.2 kb

5.1.6 Antibodies

Antibody	Species	Dilution	Origin
α-A43	rabbit	1:50000	A. Sentenac, Paris (Buhler et al., 1980)
α-A135	rabbit	1:50000	A. Sentenac, Paris (Buhler et al., 1980)
α-HA (3F10)	rat	1:5000	Roche
α-Myc (9E10)	mouse	1:200	E. Kremmer
α-mouse IgG (H+L) (peroxidase-conjugated)	goat	1:5000	Jackson IR/Dianova
α-rabbit IgG (H+L) (peroxidase-conjugated)	goat	1:5000	Jackson IR/Dianova
α-rat IgG+IgM (H+L) (peroxidase-conjugated)	goat	1:5000	Jackson IR/Dianova
PAP (peroxidase anti-peroxidase)	rabbit	1:5000	DakoCytomation

5.1.7 Enzymes

Enzyme	Origin
Antarctic Phosphatase	New England Biolabs
HotStarTaq DNA Polymerase	Qiagen
iProof High-Fidelity DNA Polymerase	Bio-Rad
Restriction Endonucleases	New England Biolabs
T4 DNA Ligase	New England Biolabs
Taq DNA Polymerase	New England Biolabs
Trypsin, sequencing grade	Roche
RNase A	Invitrogen
Proteinase K	Sigma-Aldrich

5.1.8 Kits

Kit	Origin
PureLink™ PCR Purification Kit	Invitrogen
PureLink™ Quick Gel Extraction Kit	Invitrogen
PureLink™ Quick Plasmid Miniprep Kit	Invitrogen
RadPrime DNA Labeling System	Invitrogen

5.1.9 Media

Medium	Composition
YPD (yeast extract, peptone, dextrose)	1% (w/v) yeast extract 2% (w/v) peptone 2% (w/v) glucose
YPD+gen/+hyg (YPD plus geneticin/hygromycin B)	YPD + 200 µg/ml geneticin (G418) / + 300 µg/ml hygromycin B
YPAD (YPD plus adenine)	YPD + 100 mg/l adenine
YPUD (YPD plus uracil)	YPD + 2 mg/ml uracil
YPD+dox (YPD plus doxycycline)	YPD + 0/0.02/0.1/0.5/1 µg/ml doxycycline
YPG (yeast extract, peptone, galactose)	1% (w/v) yeast extract 2% (w/v) peptone 2% (w/v) galactose
YPUG (YPG plus uracil)	YPG + 2 mg/ml uracil

MATERIAL AND METHODS

YPG+gen/+hyg (YPG plus geneticin/hygromycin B)	YPG + 200 µg/ml geneticin (G418) / + 300 µg/ml hygromycin B
SCD (synthetic complete dextrose)	0.67% (w/v) YNB + nitrogen 0.062% (w/v) CSM-his-leu-trp 2% (w/v) glucose + 20 mg/l L-histidine + 100 mg/l L-leucine + 50 mg/l L-tryptophan
SCD-leu/-trp (SCD minus leucine/tryptophan)	0.67% (w/v) YNB + nitrogen 0.062% (w/v) CSM-his-leu-trp 2% (w/v) glucose + 20 mg/l L-histidine + 50 mg/l L-tryptophan / + 100 mg/l L-leucine
SCD-his+gen (SCD minus histidine plus geneticin)	0.17% (w/v) YNB – nitrogen 0.1% (w/v) proline 0.062% (w/v) CSM-his-leu-trp 2% (w/v) glucose + 100 mg/l L-leucine + 50 mg/l L-tryptophan + 200 µg/ml geneticin (G418)
SCD-met-cys (SCD minus methionine minus cysteine)	0.67% (w/v) YNB + nitrogen 0.051% (w/v) CSM-his-leu-lys-met-trp-ura 2% (w/v) glucose + 20 mg/l L-histidine + 100 mg/l L-leucine + 50 mg/l L-lysine + 50 mg/l L-tryptophan + 20 mg/l L-uracil
SCD-leu+5FOA (SCD minus leucin plus 5FOA)	SCD-leu + 0.1% (w/v) 5FOA
SCD-trp+5FOA (SCD minus tryptophan plus 5FOA)	SCD-trp + 0.1% (w/v) 5FOA
SCG (synthetic complete galactose)	0.67% (w/v) YNB + nitrogen 0.062% (w/v) CSM-his-leu-trp 2% (w/v) galactose + 20 mg/l L-histidine + 100 mg/l L-leucine + 50 mg/l L-tryptophan
SCG-leu	0.67% (w/v) YNB + nitrogen

(SCG minus leucine)	0.062% (w/v) CSM-his-leu-trp 2% (w/v) galactose + 20 mg/l L-histidine + 50 mg/l L-tryptophan
SCR (synthetic complete raffinose)	0.67% (w/v) YNB + nitrogen 0.13% AAM-leu-ura-his + 100 mg/l L-leucine + 20 mg/l L-uracil + 20 mg/l L-histidine + 2% (w/v) raffinose
SCR-leu (SCR minus leucine)	0.67% (w/v) YNB + nitrogen 0.13% AAM-leu-ura-his + 20 mg/l L-uracil + 20 mg/l L-histidine + 2% (w/v) raffinose
SCR-leu+gal (SCR minus leucine plus galactose)	SCR-leu + 2% (w/v) galactose
LB (luria broth)	1% (w/v) tryptone 0.5% (w/v) yeast extract 0.5% (w/v) NaCl
LB+amp (LB plus ampicillin)	LB + 100 µg/ml ampicillin
SOB (super optimal broth)	2% (w/v) tryptone 0.5% (w/v) yeast extract 0.5 g/l NaCl 0.19 g/l KCl 2.03 g/l $MgCl_2 \cdot 6\,H_2O$ pH 7.0 with NaOH

AAM-leu-ura-his (amino acid mix minus leucine minus uracil minus histidine) is a mixture of the amino acids tyrosine (60 mg/l), isoleucine (80 mg/l), phenylalanine (50 mg/l), glutamic acid (100 mg/l), threonine (200 mg/l), aspartic acid (100 mg/l), valine (150 mg/l), serine (400 mg/l), arginine (20 mg/l), tryptophan (40 mg/l), methionine (20 mg/l), lysine (30 mg/l), and the nucleobase adenine (40 mg/l). Media for agar plates were supplemented with 2% (w/v) agar. All growth media were autoclaved for 20 min at 120°C. If required antibiotics, fungicides, counterselection drugs, or other chemicals were added after cooling to about 60°C.

MATERIAL AND METHODS

5.1.10 Buffers

Buffer	Ingredients	Concentration
10x PBS	NaCl	1.37 M
	KCl	27 mM
	KH_2PO_4	18 mM
	Na_2HPO_4	0.1 M
	pH 7.4 with NaOH	
10x PBST	NaCl	1.37 M
	KCl	27 mM
	KH_2PO_4	18 mM
	Na_2HPO_4	0,1 M
	Tween 20	0.5% (v/v)
	pH 7.4 with NaOH	
4x lower tris	Tris	1.5 M
	SDS	0.4% (w/v)
	pH 8.8 with HCl	
4x upper tris	Tris	0.5 M
	SDS	0.4% (w/v)
	bromophenol blue	
	pH6.8 with HCl	
4x SDS sample buffer	Tris pH 6.8	0.25 M
	glycerol	40% (v/v)
	SDS	8.4% (w/v)
	β-mercaptoethanol	0.57 M
	bromophenol blue	
HU buffer	Tris pH 6.8	0.2 M
	SDS	5% (w/v)
	EDTA pH 8.0	1 mM
	β-mercaptoethanol	0.21 M
	urea	8 M
	bromophenol blue	
10x electrophoresis buffer	Tris	0.25 M
	glycine	1.92 M
	SDS	1% (w/v)
transfer buffer (Western Blot)	Tris	25 mM
	glycine	192 mM
	methanol	20% (v/v)
ponceau staining solution	Ponceau S	0.5% (w/v)
	HOAc	1% (v/v)

MATERIAL AND METHODS

coomassie staining solution	Coomassie Brilliant Blue R-250	0.1% (w/v)
	methanol	45% (v/v)
	HOAc	10% (v/v)
destaining solution	methanol	45% (v/v)
	HOAc	10% (v/v)
100x protease inhibitors (PIs)	benzamidine	0.2 M
	PMSF	0.1 M
	solvent: ethanol	
	store at -20°C	
10x DNA loading buffer	Tris-HCl pH 8.0	4 mM
	EDTA pH 8.0	0.4 mM
	glycerol	60% (v/v)
	bromophenol blue	
	xylene cyanol	
5x TBE buffer	Tris	445 mM
	boric acid	445 mM
	EDTA pH 8.0	10 mM
hybridization buffer	sodium phosphate pH 7.2	0.5 M
	SDS	7% (w/v)
AE buffer	NaOAc pH 5.3	50 mM
	EDTA pH 8.0	10 mM
RNA solubilization buffer	formamide	50% (v/v)
	formaldehyde	8% (v/v)
	MOPS buffer	1x
	store at -20°C	
20x SSC	NaCl	3 M
	trisodium citrate dihydrate	0.3 M
	pH 7.0 with HCl	
10x MOPS buffer	sodium acetate trihydrate	20 mM
	MOPS	0.2 M
	EDTA pH 8.0	10 mM
	pH 7.0 with NaOH	
R_D buffer	glucose	2% (w/v)
	peptone	1% (w/v)
	malt extract	0.6% (w/v)
	yeast extract	0.01% (w/v)
	mannitol	12% (w/v)
	MgOAc	17.8 mM
	store at -20°C	

R_G buffer	galactose	2% (w/v)
	peptone	1% (w/v)
	malt extract	0.6% (w/v)
	yeast extract	0.01% (w/v)
	mannitol	12% (w/v)
	MgOAc	17.8 mM
	store at -20°C	
TELit	Tris pH 8.0	10 mM
	LiOAc	100 mM
	EDTA pH 8.0	1 mM
	pH 8.0 with HOAc	
LitSorb	sorbitol	1 M
	solvent: TELit	
	sterile filtration	
LitPEG	polyethylene glycol (PEG3350)	40% (w/v)
	solvent: TELit	
	sterile filtration	
rapamycin stock solution	rapamycin	1 mg/ml
	solvent: DMSO	
	store at -20°C	
cycloheximide stock solution	cycloheximide	50 mg/ml
	solvent: ethanol	
	store at -20°C	

The solvent is H_2O, if not indicated otherwise. The pH values were measured at room temperature (RT).

5.1.11 Chemicals

Chemicals were purchased at the highest purity available from Sigma-Aldrich, Merck, Fluka, Roth, or J.T.Baker, except 5FOA (Toronto Research Chemicals), electrophoresis-grade agarose (Invitrogen), bromophenol blue (Serva), G418/Geneticin (Gibco), milk powder (Sukofin), Nonidet P40 (NP40) (USB Corporation), Tris ultrapure (USB Corporation), and Tween 20 (T20) (Serva).

Ingredients for growth media were purchased from BD Becton, Dickinson and Co. (agar, peptone, tryptone and yeast extract), Qbiogene, Bio101, or Sunrise Science Products (complete supplement mixtures (CSM), yeast nitrogen base (YNB), amino acids, and adenine), Sigma-Aldrich (D(+)-glucose, D(+)-galactose, amino acids, and uracil), PerkinElmer (5',6'-[^{3}H] uracil), and Hartmann Analytic (α-[^{32}P]-ATP, [^{35}S]-met/cys). Water was always purified with an Elga Purelab Ultra device prior to use.

MATERIAL AND METHODS

5.1.12 Other materials

Material	Origin
Protein Marker, Broad Range (2-212 kDa)	New England Biolabs
ColorPlus Prestained Protein Marker, Broad Range (7-175 kDa)	New England Biolabs
1 kb DNA ladder	New England Biolabs
100 bp DNA ladder	New England Biolabs
Salmon Sperm DNA (10 mg/ml)	Invitrogen
yeast genomic DNA (strain S288C)	Invitrogen
Immobilon-P Membrane PVDF 0,45 µm	Millipore
Membrane Positive™	MP Biomedicals
Blotting papers MN 827 B	Millipore
Extra Thick Blot Paper	Bio-Rad
BM Chemiluminescence Blotting Substrate (POD)	Roche
SimplyBlue™ SafeStain	Invitrogen
IgG Sepharose™ 6 Fast Flow	GE Healthcare
Protein G Sepharose™ 4 Fast Flow	GE Healthcare
Protein Assay Dye Reagent Concentrate	Bio-Rad
SYBR Safe DNA Gel Stain	Invitrogen
SYBR Green	Roche
glass beads (∅ 0.75-1 mm)	Roth
Calmodulin Affinity Resin	Stratagene
Glutathione Sepharose 4B	GE Healthcare
BioMax MS Film	Sigma-Aldrich
EN³HANCE Spray Surface Autoradiography Enhancer	PerkinElmer

5.1.13 Equipment

Device	Manufacturer
4700 Proteomics Analyzer MALDI-TOF/TOF	Applied Biosystems
4800 Proteomics Analyzer MALDI-TOF/TOF	Applied Biosystems
Biofuge Fresco refrigerated tabletop centrifuge	Hereaus
Biofuge Pico tabletop centrifuge	Hereaus
C412 centrifuge	Jouan
Centrikon T-324 centrifuge	Kontron Instruments
CT422 refrigerated centrifuge	Jouan
Electrophoresis system model 45-2010-i	Peqlab Biotechnologie GmbH
FastPrep Instrument	Qbiogene
FPLC-System (Pumps P-500; Controller LCC-501+; Fraction	Pharmacia Biotech

MATERIAL AND METHODS

Collector FRAC-100)	
Gel Max UV transilluminator	Intas
IKA-Vibrax VXR	IKA
Incubators	Memmert
LAS-3000 chemiluminescence imager	Fujifilm
MicroPulser electroporation apparatus	Bio-Rad
NanoDrop ND-1000 spectrophotometer	Peqlab Biotechnologie GmbH
Avanti J-20 XP centrifuge	Beckman Coulter
Optima L-80 X ultracentrifuge	Beckman Coulter
PCR Sprint thermocycler	Hybaid
Power Pac 3000 power supplies	Bio-Rad
Pulverisette 6 planetary mono mill	Fritsch
Roto-Shake Genie	Scientific Industries
Shake incubators Multitron / Minitron	Infors
Speed Vac Concentrator	Savant
Thermomixer compact	Eppendorf
Trans-Blot SD Semi-Dry Transfer Cell	Bio-Rad
Ultrospec 3100pro spectrophotometer	Amersham
XCell SureLock Mini-Cell electrophoresis system	Invitrogen
AxioCam MR CCD camera	Zeiss
Axiovert 200M microscope	Zeiss
Rotor-Gene 3000	Corbett Research
Sonifier 250	Branson
SMART System	Pharmacia Biotech
Mono Q PC 1.6/5	Pharmacia Biotech
Superose 12 PC 3.2/30	GE Healthcare
FLA-3000 phosphor imager	Fujifilm

5.1.14 Software

Software	Producer
4000 Series Explorer v.3.6	Applied Biosystems
Acrobat 7.0 Professional v.7.0.9	Adobe
Data Explorer v.4.5 C	Applied Biosystems
GPS Explorer v.3.5	Applied Biosystems
Image Reader LAS-3000 V2.2	Fujifilm
Mascot	Matrix Science
Microsoft Office 2007	Microsoft

MATERIAL AND METHODS

ND-1000 v.3.5.2	Peqlab Biotechnologie GmbH
Photoshop CS v.8.0.1	Adobe
Axiovision V 4.7.1.0	Zeiss
Multi Gauge V3.0	Fujifilm
Rotor-Gene V6.1	Corbett Research
Image Reader FLA-3000 V1.8	Fujifilm

5.2 Methods

5.2.1 Work with *Saccharomyces cerevisiae*

5.2.1.1 Cultivation of yeast strains

Strains of the yeast *Saccharomyces cerevisiae* were cultivated using standard microbiological methods (Sherman, 2002).

Liquid cultures were grown in the appropriate medium usually at 30°C, except for temperature-sensitive mutants (24°C) or for temperature-shift experiments (24°C, 37°C). Cell growth was monitored by measuring the optical density at 600 nm (OD_{600}).

For cultivation on solid agar plates containing the appropriate medium, single colonies or small aliquots of glycerol stocks were streaked out using sterile disposable inoculation loops in order to obtain colonies derived from single yeast cells. Plates were incubated upside down at the respective temperatures for 1-5 days. Short-term storage of yeast strains was accomplished by keeping the agar plates at 4°C.

5.2.1.2 Preparation of competent yeast cells

50 ml of a logarithmically growing yeast culture (OD_{600} ~ 0.5-0.7) were harvested by centrifugation for 5 min at 4000 rpm and RT. Cells were washed with 25 ml sterile H_2O and 5 ml LitSorb before resuspending in 360 µl LitSorb. 40 µl of Salmon Sperm DNA, which were incubated for 5 min at 99°C and immediately chilled on ice, were added to the cell suspension. After mixing, 50 µl aliquots were transferred to 1.5 ml tubes before storage at -80°C.

5.2.1.3 Transformation of competent yeast cells

The aliquot containing competent yeast cells was thawed on ice. DNA to be transformed (100 ng of plasmid DNA or 5-10 µg of linear DNA, respectively) was added and the sample was mixed. After addition of six volumes of LitPEG, the suspension was again mixed thoroughly and incubated for 30 min at RT on a turning wheel. The sample was mixed with 1/9 total volume (cells plus DNA plus LitPEG) of sterile DMSO followed by a heat-shock at 42°C for 15 min and centrifugation for 1 min at 3000 rpm and RT.

When selecting for auxotrophic markers (e.g. TRP1, LEU2, HIS3, URA3), the cell pellet was directly resuspended in 100 µl sterile H$_2$O and plated on SCD- or SCG-plates, respectively, lacking the corresponding amino acid.

If selection for antibiotic resistance (e.g. geneticin, hygromycin B) was required, the cell pellet was resuspended in 5 ml YPD or YPG, respectively, and incubated for 3-6 h at 30°C while shaking to allow the expression of the marker before plating on the respective selection media. Since selection for antibiotic resistance often results in a high number of transient transformants, these plates were replica-plated on fresh selection media to identify positive clones.

5.2.1.4 Crosslinking of yeast cells with formaldehyde

Logarithmically growing cells from 45 ml liquid culture were crosslinked by adding 1.25 ml 37% (v/v) formaldehyde and subsequent incubation for additional 15 min at the respective growth temperature while shaking. Crosslinking was quenched by adding 2.5 ml 2.5 M glycine. The culture was further incubated at growth temperature for additional 5 min. Cells were harvested by centrifugation in a 50 ml tube for 3 min at 3000 rpm and 4°C. The cell pellet was washed with cold 1x PBS, transferred to a 1.5 ml tube, and frozen in liquid nitrogen before storage at -20°C.

5.2.1.5 Yeast plasmid shuffle

The plasmid shuffle yeast strains pRRN3 (URA) (Y2189) and pRRN3 (URA)-A43-HA$_3$ (Y2113) were used to replace the essential *RRN3* gene by mutant alleles. In these strains the chromosomal locus of *RRN3* is replaced by the *HIS3* marker gene and the deletion is complemented by a wild type copy of *RRN3* on a plasmid containing the *URA3* marker gene. The shuffle strains were transformed with another plasmid carrying the mutant allele of *RRN3* and plated on the corresponding selection medium. If the mutant allele is able to complement the chromosomal deletion, the plasmid containing the wild type copy of *RRN3* can be lost during cultivation. Loss of the wild type allele of *RRN3* on the plasmid containing the marker gene *URA3* was monitored by growing the cells on SCD+5FOA (5-fluoroorotic acid) plates lacking the corresponding amino acid, since growth on this counterselection medium is lethal for all cells still containing the wild type allele of *RRN3*. *URA3* codes for the enzyme orotidin-5'-phosphate decarboxylase of the uracil-biosynthesis pathway, which converts 5FOA into the toxic 5-fluorouracil.

In each case, a transformation of the shuffle strains with an appropriate plasmid not carrying an *RRN3* allele was performed. Single clones from these control transformations were supposed to grow on the corresponding selection medium but not on the corresponding counterselection medium as *RRN3* is an essential gene.

Putative positive clones were finally monitored for the presence of the mutant plasmid, the loss of the wild type plasmid, and the maintenance of the chromosomal deletion by analyzing the respective auxotrophic markers.

5.2.1.6 Spot test analysis of yeast strains

Overnight cultures of the yeast strains to be tested were diluted to OD_{600} = 0.1 with sterile H_2O. 7-10 µl of this cell suspension and of serial 1:10, 1:100, and 1:1000 dilutions with sterile H_2O were spotted on the appropriate test plates. Phenotypes were monitored after incubation for 2-4 days at 16°C, 24°C, 30°C, and 37°C, respectively.

5.2.1.7 Growth kinetic analysis of yeast strains

Cultures of the yeast strains to be tested were grown to stationary phase overnight. From these pre-cultures fresh cultures were inoculated to OD_{600} = 0.1 in the appropriate medium and growth at 30°C was monitored by measuring the OD_{600} mostly in 1 h intervals. Since the strains were always kept in the logarithmic growth phase (OD_{600} ~ 0.1-0.7) by dilution with the respective medium, the dilution factor had to be taken into account for the calculation of the growth kinetics.

5.2.1.8 Long-term storage of yeast strains

2 ml of an overnight culture of the yeast strain to be stored were mixed with 1 ml sterile 50% (v/v) glycerol and separated into two aliquots. Glycerol stocks were stored at -80°C.

5.2.2 Work with *Escherichia coli*

5.2.2.1 Cultivation of bacterial strains

Liquid cultures were grown in LB(+amp) medium at 37°C. Cell growth was monitored by measuring the optical density at 600 nm (OD_{600}). For cultivation on solid agar plates containing LB(+amp) medium, single colonies or small aliquots of glycerol stocks were streaked out using sterile disposable inoculation loops in order to obtain colonies derived from single bacterial cells. Plates were incubated upside down at 37°C for 1 day. Short-term storage of bacterial strains was accomplished by keeping the agar plates at 4°C.

5.2.2.2 Preparation of competent bacterial cells for electroporation

The XL1-Blue strain was used as a host for amplification of plasmid DNA. In order to increase the efficiency of plasmid DNA uptake, competent cells for electroporation were prepared. Cells were grown in 400 ml SOB medium at 37°C to mid-log phase (OD_{600} ~ 0.35-0.6), chilled on ice for 15 min, and centrifuged for 10 min at 6000 rpm and 4°C. To reduce the ionic strength of the cell suspension, cells were washed 3x with cold sterile H_2O and 1x with sterile 10% (v/v) glycerol. After resuspending the cells in 1.5 ml sterile 10% (v/v) glycerol, 50 µl aliquots were transferred to 1.5 ml tubes before storage at -80°C.

5.2.2.3 Transformation of competent bacterial cells by electroporation

The aliquot containing competent bacterial cells was thawed on ice. DNA to be transformed (1 ng of plasmid DNA or up to 3 µl of a ligation sample) was added and the sample was mixed. After pipetting the suspension into a cold 0.2 cm electroporation cuvette, pulsing was performed with program EC2 in a MicroPulser electroporation apparatus. Immediately after the pulse, 1 ml LB medium was added and the sample was transferred to a 1.5 ml tube following incubation for 30-60 min at 37°C. 100 µl of the cell suspension were plated on LB+amp and incubated overnight at 37°C. The residual cells were centrifuged for 1 min at 5000 rpm and RT. About 900 µl of the supernatant were discarded and the pellet was resuspended in the remaining liquid, plated on LB+amp, and incubated overnight at 37°C.

5.2.3 Work with DNA

5.2.3.1 Phenol-chloroform extraction

To separate DNA in an aqueous solution from proteins and RNA, one volume of a phenol:chloroform:isoamyl alcohol-mixture (25:24:1) was added to the sample. The samples were mixed by vortexing until the solution was milky. After centrifugation for 5 min at 13000 rpm and RT, an aliquot of the upper aqueous phase was transferred to a 1.5 ml tube without disturbing the white layer of denatured protein between the upper aqueous and the lower phenol phase.

5.2.3.2 Purification of DNA by ethanol precipitation

DNA was precipitated from aqueous solution by mixing the sample with 1/10 volume of 3 M NaOAc pH 5.3 and 2.5 volumes of ethanol following incubation for at least 20 min at -20°C. Ethanol depletes the hydration shell from nucleic acids and expose negatively charged phosphate groups. Counter cations (here Na^+) bind the charged groups and reduce the repulsive forces between the polynucleotide chains, allowing the formation of a precipitate. Samples were centrifuged for 20 min at 13000 rpm and 4°C. To eliminate salt, the pellet was washed with cold 70% (v/v) ethanol. After removal of the supernatant, the nucleic acid pellet was dried at RT and solubilized in an appropriate volume of TE buffer [10 mM Tris-HCl pH 8.0, 1 mM EDTA pH 8.0].

5.2.3.3 DNA quantification using UV spectroscopy

Concentration of pure DNA samples was measured by UV spectroscopy at 260 nm using a NanoDrop ND-1000 spectrophotometer (1 OD_{260} = 50 µg/ml). To determine contamination with proteins, absorbance was concomitantly measured at 280 nm. The ratio of OD_{260}/OD_{280} of pure DNA is between 1.8 and 2.0.

MATERIAL AND METHODS

5.2.3.4 Native agarose gel electrophoresis

Native agarose gel electrophoresis was used to separate DNA fragments of different lengths. In this work, electrophoresis was performed routinely with gels composed of 1.0-1.2% (w/v) agarose and 1x TBE and containing 0.5 µg/ml ethidium bromide (EtBr) or 1x SYBR Safe DNA Gel Stain, respectively. 1x TBE was used as electrophoresis buffer and gels were run at 100-120 V. To determine the lengths of the fragments, 0.5 µg of a DNA standard (1 kb ladder or 100 bp ladder) was used in a concentration of 50 µg/ml in 2.5x DNA loading buffer. DNA fragments could be visualized by exposing the gel to UV light (470 nm).

5.2.3.5 Southern Blot (passive capillary transfer)

Separated DNA fragments were transferred and immobilized on a positively charged membrane using the passive capillary transfer method. Prior to the transfer, the agarose gel was incubated for 2x 15 min in 0.5 M NaOH/1.5 M NaCl on a shaker to denature double-stranded DNA. Subsequently, the gel was incubated for 2x 15 min in 1 M NH_4OAc. Transfer of the DNA fragments from the agarose gel to the membrane was then achieved overnight by drawing the transfer buffer (1 M NH_4OAc) from the reservoir upward through the gel and the membrane into a stack of paper towels. The DNA fragments were eluted from the gel and deposited onto the positively charged membrane with the help of the buffer stream. After the transfer, the DNA fragments were crosslinked to the dried membrane by exposition for 20 sec to UV light (0.3 J/cm^2).

5.2.3.6 Radioactive probe labeling, hybridization, and detection

Probes were radioactively labeled with α-[^{32}P]-ATP using the RadPrime DNA Labeling System as indicated by the manufacturer.

The membranes, separated by meshes in a hybridization tube, were pre-hybridized for 1 h at hybridization temperature (65°C) with 50 ml hybridization buffer. For hybridization, the membranes were incubated in fresh 15 ml hybridization buffer. The labeled probe was mixed with 150 µl Salmon Sperm DNA, incubated for 5 min at 99°C, immediately chilled on ice, and added to the hybridization tube. Hybridization occurred overnight at hybridization temperature with tubes rotating in a hybridization oven. The membranes were first rinsed 1x with 30 ml 3x SSC/0.1% (w/v) SDS and subsequently washed in each case for 2x 15 min with 30 ml 0.3x SSC/0.1% (w/v) SDS, 0.1x SSC/0.1% (w/v) SDS and 0.1x SSC/1.5% (w/v) SDS at hybridization temperature. Afterwards the membranes were dried and covered with an erased phosphor-imaging plate (PIP) in a cassette. The time of exposure depended on the intensity of the radioactive signal. The PIP was scanned with a resolution of 100 µm by a FLA-3000 phosphor imager using Image Reader FLA-3000 V1.8 followed by quantitative analysis using Multi Gauge V3.0.

MATERIAL AND METHODS

5.2.3.7 Polymerase Chain Reaction (PCR)

For amplification of DNA fragments for integration in the yeast genome, PCR was performed with yeast genomic DNA or plasmid DNA (50-100 ng) as templates in five 100 µl reactions [20 mM Tris-HCl pH 8.8, 10 mM $(NH_4)_2SO_4$, 10 mM KCl, 2 mM $MgSO_4$, 0.1% (v/v) Triton X-100, 0.25 µM of forward and reverse primers, 0.25 mM dNTPs, 2.5-5 U Taq Polymerase]. The main PCR program used in this work was as following:

95°C	5 min	(1x)
95°C	1 min	
45°C	1 min	(35x)
72°C	2 min	
72°C	10 min	(1x)

For amplification of DNA fragments used for cloning, proofreading PCR was performed with yeast genomic DNA or plasmid DNA (50-100 ng) as templates in a 100 µl reaction [1x iProof HF Buffer, 0.5 µM of forward and reverse primers, 0.2 mM dNTPs, 2 U iProof High-Fidelity DNA Polymerase]. The PCR program was designed as indicated by the manufacturer.

10% of the reactions were analyzed by native agarose gel electrophoresis and subsequently purified with a PCR Purification Kit according to the manufacturer.

5.2.3.8 Quantitative real-time Polymerase Chain Reaction (qPCR)

Quantitative real-time PCR was used to measure the amount of a specific DNA fragment with high accuracy. The amount of DNA present at the end of each single PCR cycle was detected by measuring the fluorescence of SYBR Green. SYBR Green is a dye that exhibits fluorescence when bound to double-stranded DNA but not in solution. Therefore, the intensity of the fluorescence signal allows direct measurement of the actual amount of DNA present in the sample. Quantitative real-time PCR reactions were performed in 0.1 ml tubes, the reaction volume was 20 µl. The reaction contained 4 µl of DNA sample and 16 µl of master mix. The master mix contained 4 pmol of both the forward and the reverse primer, 0.25 µl of a 1:400000 SYBR Green stock solution in DMSO, 0.4 U HotStarTaq DNA Polymerase and a premix. The premix was composed of $MgCl_2$ (final concentration in the reaction: 2.5 mM), dNTPs (final concentration in the reaction: 0.2 mM) and 10x PCR buffer (final concentration in the reaction: 1x). Quantitative real-time PCR was performed in a Rotor-Gene 3000. The data were analyzed with Rotor-Gene V6.1.

5.2.3.9 Digestion of DNA with restriction endonucleases

A variety of prokaryotic restriction endonucleases were used in this work to digest DNA in order to prepare defined DNA fragments for cloning or to check for presence and correct orientation of inserted DNA fragments. Restriction endonucleases were essentially used as suggested by the manufacturer.

MATERIAL AND METHODS

5.2.3.10 DNA ligation
In order to clone DNA sequences into yeast/bacterial plasmids, the quantity of purified DNA fragments digested with restriction endonucleases was measured by UV spectroscopy (see section 5.2.3.3). A threefold molar excess of insert DNA compared to the plasmid DNA fragment was incubated in a 20 µl ligase reaction (50 mM Tris-HCl pH 7.5, 10 mM $MgCl_2$, 1 mM ATP, 10 mM DTT, 400 U T4 DNA Ligase) for 2 h at RT or overnight at 16°C. Up to 3 µl of the ligation reaction were used for transformation of competent bacterial cells (see section 5.2.2.3).

5.2.3.11 DNA sequencing and oligonucleotide synthesis
DNA sequencing was performed by GENEART and the service of primer synthesis was provided by Eurofins MWG Operon. Oligonucleotides used in this work are listed in section 5.1.4.

5.2.3.12 Plasmid construction

5.2.3.12.1 pRRN3-Prot.A
Both the DNA fragment derived from PCR reaction (yeast genomic DNA/2234/2235) and the plasmid pNOP1-RRN3-Prot.A were digested with the restriction endonucleases SacI and NdeI. The plasmid pRRN3-Prot.A was generated by ligation of the respective DNA fragment into the backbone of the plasmid.

5.2.3.12.2 pGAL1-RRN3-Prot.A (A/C) and pGAL1-RRN3-Prot.A (2µ)
Both the DNA fragment derived from PCR reaction (pRRN3-Prot.A/2472/2473) and the plasmids pGAL1-RRN3 (A/C) and pGAL1-RRN3 (2µ) were digested with the restriction endonucleases NdeI and HindIII. The plasmids pGAL1-RRN3-Prot.A (A/C) and pGAL1-RRN3-Prot.A (2µ) were generated by ligation of the respective DNA fragment into the backbone of the plasmids.

5.2.3.12.3 pRS315-RRN3 (S102A) and pRS315-RRN3 (S102D)
In the first case, two PCR reactions (pRS315-RRN3/2141/2137 and pRS315-RRN3/2139/2142) were performed to generate two appropriate DNA fragments which were both used as templates in an additional PCR reaction (2141/2142). Both the DNA fragment derived from this PCR reaction and the plasmid pRS315-RRN3 were digested with the restriction endonucleases HindIII and NdeI. The plasmid pRS315-RRN3 (S102A) was generated by ligation of the respective DNA fragment into the backbone of the plasmid.
In the second case, two PCR reactions (pRS315-RRN3/2141/2138 and pRS315-RRN3/2140/2142) were performed to generate two appropriate DNA fragments which were both used as templates in an additional PCR reaction (2141/2142). Both the DNA fragment derived from this PCR reaction and the plasmid pRS315-RRN3 were digested with the restriction endonucleases

HindIII and NdeI. The plasmid pRS315-RRN3 (S102D) was generated by ligation of the respective DNA fragment into the backbone of the plasmid.

5.2.4 Work with RNA

5.2.4.1 RNA extraction

RNA extractions were essentially performed as described previously (Schmitt et al., 1990). Cell pellets were resuspended in 500 µl AE buffer and mixed with 500 µl phenol equilibrated in AE buffer and 50 µl of 10% (w/v) SDS. The samples were incubated in a thermomixer for 5 min at 1400 rpm and 65°C and afterwards chilled on ice for 2 min. After centrifugation for 2 min at 13000 rpm and RT, 3x 150 µl of the aqueous phase was collected and mixed with 500 µl phenol equilibrated in AE buffer by vortexing. The samples were again centrifuged and 3x 120 µl of the supernatant were mixed with 500 µl chloroform by vortexing. Phases were separated by centrifugation and the RNA in 3x 100 µl of the supernatant was precipitated by addition of 1/10 volume of 3 M NaOAc pH 5.3 and 2.5 volumes of ethanol and incubation for 30 min at -20°C. The precipitated RNA, when used for denaturing agarose gel electrophoresis, was solubilized in RNA solubilization buffer, denatured by incubation for 15 min at 65°C, and stored at -20°C.

5.2.4.2 Denaturing agarose gel electrophoresis

Denaturing agarose gel electrophoresis was used to separate RNA species longer than 1000 bases. In this work electrophoresis was performed routinely with gels composed of 1.3% (w/v) agarose, 2% (v/v) formaldehyde, and 1x MOPS containing 0.5 µg/ml EtBr. The electrophoresis buffer was composed of 1x MOPS and 2% (v/v) formaldehyde. Gels were run for 14-16 h at 40 V.

5.2.4.3 Northern Blot (passive capillary transfer)

Separated [^{3}H]-labeled RNAs were transferred and immobilized on a positively charged membrane using the passive capillary transfer method. Prior to the transfer, the agarose gel was washed for 5 min in H$_2$O, for 20 min in 50 mM NaOH to hydrolyze the RNAs and facilitate the transfer of larger RNAs, and was further equilibrated for 2x 20 min in 10x SSC. Transfer of the RNAs from the agarose gel to the membrane was then achieved overnight by drawing the transfer buffer (10x SSC) from the reservoir upward through the gel and the membrane into a stack of paper towels. The RNAs were eluted from the gel and deposited onto the positively charged membrane with the help of the buffer stream. After the transfer, the RNAs were crosslinked to the dried membrane by exposition for 1 min to UV light (254 nm/312 nm).

MATERIAL AND METHODS

5.2.4.4 Detection of [^{3}H]-labeled RNAs

The membrane was covered with an erased tritium-imaging plate (TIP) in a cassette. The time of exposure depended on the intensity of the radioactive signal. The TIP was scanned with a resolution of 100 µm by a FLA-3000 phosphor imager using Image Reader FLA-3000 V1.8 followed by quantitative analysis using Multi Gauge V3.0.

Alternatively or subsequently, the membrane was sprayed with EN^{3}HANCE solution and subjected to autoradiography.

5.2.4.5 Analysis of neo-synthesized RNA

In one case, cells were grown in YPD medium at 30°C and further cultivated either in the presence or in the absence of rapamycin or cycloheximide, respectively. In the other case, cells were grown in SCR medium depleted of leucine at 30°C before galactose was added and cells were shifted for 8 h to 24°C.

For each sample 1 OD_{600} of cells was centrifuged for 1 min at 10000 rpm and RT before the cell pellets were resuspended in 100 µl buffer R_D (cells grown in media based on glucose) or R_G (cells grown in media based on galactose), respectively. 60 µCi or 20 µCi of 5',6'-[^{3}H]-uracil were added and the cells were incubated either for 5 min at 30°C or 20 min at 24°C (pulse), before one volume of either YPUD or YPUG was added and cells were incubated either for additional 4 min, 8 min and 16 min at 30°C or for additional 60 min at 24°C (chase). Immediately after the treatment, the samples were chilled on ice and centrifuged for 1 min at 13000 rpm and 4°C. The supernatants were discarded and the cell pellets were stored at -20°C. Total RNA was extracted as described (see section 5.2.4.1), same amounts of samples were separated by denaturing agarose gel electrophoresis as described (see section 5.2.4.2), transferred to a membrane as described (see section 5.2.4.3), and analyzed as described (see section 5.2.4.4).

5.2.5 Work with proteins

5.2.5.1 Preparation of yeast whole cell extract (WCE)

5.2.5.1.1 Small scale

For whole cell extract preparation in a small scale, logarithmically growing cells from 50 ml liquid culture of OD_{600} ~ 0.8 were harvested by centrifugation in a 50 ml tube for 3 min at 3000 rpm and 4°C. The cells were washed with 1 ml cold H_2O and transferred to a 1.5 ml tube. After resuspending the cell pellet in an equal volume of high-salt extraction buffer [150 mM HEPES pH 7.6, 400 mM $(NH_4)_2SO_4$, 10 mM $MgCl_2$, 20% (v/v) glycerol, 5 mM β-mercaptoethanol, 2 mM benzamidine, 1 mM PMSF], an equal volume of cold glass beads (∅ 0.75-1 mm) was added to the suspension and the tube was vigorously shaken either on an IKA-Vibrax VXR for 3x 15 min at 2200 rpm and 4°C or in a FastPrep Instrument for 2x 40 sec at 5.5 m/s and 4°C. The tube was pierced on the bottom and on the lid with a hot needle and placed in a 15 ml tube to remove

MATERIAL AND METHODS

the glass beads from the cell extract by centrifugation for 2 min at 2000 rpm and 4°C. The cell extract was transferred to a 1.5 ml tube and centrifuged again for 1 min at 13000 rpm and 4°C to remove cell debris. The cleared supernatant (WCE) was transferred to a 1.5 ml tube and appropriate amounts were supplemented with 4x SDS sample buffer, incubated for 5 min at 99°C and analyzed as described (see sections 5.2.5.6, 5.2.5.7, and 5.2.5.8) or if required frozen in liquid nitrogen before storage at -80°C. Protein concentration was determined using the Bradford assay (Bradford, 1976).

5.2.5.1.2 Large scale

For whole cell extract preparation in a large scale, logarithmically growing cells from 10 l liquid culture of OD_{600} ~ 0.8 were harvested by centrifugation in 1 l tubes for 8 min at 6000 rpm and 4°C. The cells were washed with cold H_2O, pooled, and transferred to a 50 ml tube. After resuspending the cell pellet in an equal volume of high-salt extraction buffer [150 mM HEPES pH 7.6, 400 mM $(NH_4)_2SO_4$, 10 mM $MgCl_2$, 20% (v/v) glycerol, 5 mM β-mercaptoethanol, 2 mM benzamidine, 1 mM PMSF], an equal volume of cold glass beads (∅ 0.75-1 mm) was added to the suspension and the mixture was shaken in a Pulverisette 6 planetary mono mill at 500 rpm and 4°C for 3x 4 min. Glass beads were separated by filtration prior to centrifugation of the cell extract in 50 ml tubes for 20 min at 4500 rpm and 4°C to remove cell debris. The supernatant was additionally ultracentrifuged for 90 min at 100000 g and 4°C, the clear middle phase (WCE) was separated from the turbid insoluble matter in the lower phase and the top layer of lipids and subjected to protein purification (see section 5.2.6.4) or if required frozen in liquid nitrogen before storage at -80°C. Protein concentration was determined using the Bradford assay (Bradford, 1976).

5.2.5.2 Determination of protein concentration

Protein concentrations were determined using the Bio-Rad Protein Assay which is based on the method by Bradford (Bradford, 1976). Briefly, 1-5 µl of the protein solution to be tested were mixed with 1 ml protein assay dye after diluting the reagent to the working concentration according to the instructions of the manufacturer. The approximate protein concentrations in µg/µl were calculated by dividing the absorbance at 595 nm by the sample volume and multiplying with the factor 23 which was determined using a BSA standard curve.

5.2.5.3 TCA precipitation

The volume of the protein sample to be analyzed was adjusted to 100 µl with cold H_2O prior to mixing with 10 µl cold 100% (w/v) TCA and 2 µl 2% (w/v) DOC (Bensadoun and Weinstein, 1976). After incubation for 30 min at 4°C, the precipitated proteins were pelleted by centrifugation for 15 min at 13000 rpm and 4°C. The supernatant was discarded and the pellet was solubilized in an adequate volume of SDS sample buffer. The pH of the sample was neutralized using NH_3 gas, if necessary. Proteins were denatured by incubating the sample for 5 min at 99°C for subsequent separation by SDS-PAGE.

5.2.5.4 Methanol-chloroform precipitation

Protein precipitation for subsequent mass spectrometric analyses was performed using the methanol-chloroform precipitation method (Wessel and Flügge, 1984). The volume of the sample was adjusted to 150 µl with H_2O, followed by the addition of four volumes of methanol (600 µl), one volume of chloroform (150 µl), and three volumes of H_2O (450 µl). After each of these steps the sample was mixed well by vortexing. After incubation for 5 min at 4°C, the sample was centrifuged for 5 min at 13000 rpm and 4°C. The supernatant was discarded without disturbing the interphase which contains the precipitated proteins. Upon addition of another three volumes of methanol (450 µl) and vortexing, the sample was incubated for 5 min at 4°C before centrifugation for 5 min at 13000 rpm and 4°C. The supernatant was completely removed and the protein pellet dried for 10 min in a Speed Vac Concentrator.

5.2.5.5 Denaturing protein extraction

Cell pellets were resuspended in 1 ml cold H_2O, mixed with 150 µl pre-treatment solution [1.85 M NaOH, 1 M β-mercaptoethanol], and incubated for 15 min at 4°C. Proteins were precipitated with 150 µl 55% (w/v) TCA for 15 min at 4°C and pelleted by centrifugation for 10 min at 13000 rpm and 4°C. The supernatant was discarded and the pellet was resuspended in 25 µl HU buffer. The pH of the sample was neutralized using NH_3 gas, if necessary. Proteins were denatured by incubating the sample for 10 min at 65°C for subsequent separation by SDS-PAGE.

5.2.5.6 SDS-polyacrylamide gel electrophoresis (SDS-PAGE)

Proteins were separated according to their molecular weight using the vertical discontinuous SDS-polyacrylamide gel electrophoresis method by Laemmli (Laemmli, 1970). The discontinuous system consists of a lower separating gel composed of 8-12% acrylamide, 375 mM Tris-HCl pH 8.8, and 0.1% (w/v) SDS and an upper stacking gel composed of 4% acrylamide, 125 mM Tris-HCl pH 6.8, and 0.1% (w/v) SDS.

Ribosomal proteins were separated according to their low molecular weight using acrylamide urea gels. In this case, the lower separating gel is composed of 16% acrylamide, 375 mM Tris-HCl pH 8.8, 0.1% (w/v) SDS, and 4.5 M urea and the upper stacking gel is composed of 4% acrylamide, 125 mM Tris-HCl pH 6.8, 0.1% (w/v) SDS, and 4.5 M urea.

Gels were run for 1.5-2.5 h at 50 mA and 180 V in 1x electrophoresis buffer. Molecular weights of the different proteins were estimated using protein markers of known molecular weight.

MATERIAL AND METHODS

5.2.5.7 Western Blot
Separated proteins were transferred to a PVDF membrane using a Trans-Blot SD Semi-Dry Transfer Cell. The gel and the PVDF membrane, pretreated with methanol, were placed in the transfer cell between two piles of three blotting papers soaked with transfer buffer. Transfer was performed for 1 h at 24 V.

To control the blotting of the proteins before immunodetection, the total protein content was reversibly stained with Ponceau S by incubating the membrane in ponceau staining solution for 2 min and subsequent destaining with H_2O until the protein bands were visible.

5.2.5.8 Detection of proteins by chemiluminescence
Prior to specific immunodetection of defined proteins, the membrane was blocked with non-related proteins from bovine milk to avoid unspecific binding of the antibodies by incubating the membrane in 5% (w/v) milk powder in 1x PBS for 1 h at RT or overnight at 4°C on a shaker. The antibodies were diluted to an adequate working concentration in 5% (w/v) milk powder in 1x PBST. The incubations were performed in small bags made of sealed plastic foils on a turning wheel for 1 h (primary antibodies) or 30 min (secondary antibodies) at RT. Following each antibody incubation step, the membrane was washed in 1x PBST for 2x 10 min on a shaker.

In order to detect the specifically bound antibodies, the membrane was incubated for 1 min at RT with 2-4 ml BM Chemiluminescence Blotting Substrate (POD) which was prepared according to the instructions of the manufacturer. This reagent contains hydrogen peroxide and luminol which is a substrate for the horseradish peroxidase conjugated to the secondary antibodies used in this work. The light, which is emitted during this reaction at the corresponding specific positions on the membrane, was detected with a LAS-3000 chemiluminescence imager using Image Reader LAS-3000 V2.2 followed by quantitative analysis using Multi Gauge V3.0.

5.2.5.9 Coomassie staining
The polyacrylamide gel was stained with Coomassie Brilliant Blue R-250 in order to visualize the total protein content.

Briefly, the gel was stained in coomassie staining solution for 30 min on a shaker at RT. Destaining was performed by incubating the gel in destaining solution for 3-4x 30 min until protein bands showed up significantly over the background staining. Optionally, the gel could be dried in a vacuum gel dryer system for 2 h at 80°C or bands could be excised for subsequent protein identification using mass spectrometry (see section 5.2.5.10).

5.2.5.10 Protein identification using MALDI-TOF/TOF mass spectrometry
Protein bands of interest were excised from coomassie-stained gels and digested in gel with modified sequencing grade trypsin (Shevchenko et al., 1996, 2006). Briefly, the excised pieces were cut into small cubes and subsequently washed with 50 mM NH_4HCO_3, 50 mM

MATERIAL AND METHODS

NH_4HCO_3/25% (v/v) acetonitrile, 25% (v/v) actetonitrile, and 50% (v/v) acetonitrile followed by lyophilization. The dried gel cubes were rehydrated with an equal volume of trypsin in 50 mM NH_4HCO_3 (final concentration: 4 µg trypsin per 100 µl gel) for 30 min at RT. After addition of another volume of 50 mM NH_4HCO_3 and incubation for 16 h at 37°C, the resulting tryptic peptides were eluted by diffusion upon shaking the gel cubes 2x for 1 h in two volumes of 100 mM NH_4HCO_3 and 1x for 1 h in two volumes of 100 mM NH_4HCO_3/35% acetonitrile at 37°C. The supernatants of these elution steps were pooled in a 1.5 ml tube and the solvents removed by lyophilization. The extracted peptides were solubilized in 5 µl matrix solution [2 mg/ml α-cyano-4-hydroxycinnamic acid (CHCA), 50% (v/v) acetonitrile, 0,1% (v/v) TFA] and manually spotted on the MALDI target plate.

Peptide mass fingerprints (PMF) and MS/MS analyses were performed on a 4700 or a 4800 Proteomics Analyzer MALDI-TOF/TOF mass spectrometer operated in positive ion reflector mode and evaluated by searching the NCBInr protein sequence database with Mascot implemented in GPS Explorer v.3.5.

5.2.5.11 Analysis of neo-synthezised protein

Cells were grown in SCD medium depleted of methionine and cysteine at 30°C and further cultivated either in the absence or in the presence of rapamycin or cycloheximide, respectively. For each sample 2 OD_{600} of cells were centrifuged for 1 min at 10000 rpm and RT before the cell pellets were resuspended in 200 µl SCD-met-cys. 15 µCi of [^{35}S]-met/cys were added and the cells were incubated for 5 min at 30°C (pulse). Immediately after the treatment, the samples were centrifuged for 1 min at 13000 rpm and 4°C. The supernatants were discarded and the cell pellets were stored at -20°C. Total protein was extracted as described (see section 5.2.5.5), same amounts of samples were separated by SDS-polyacrylamide gel electrophoresis as described (see section 5.2.5.6), and coomassie staining was performed as described (see section 5.2.5.9). The dried gel was subjected to autoradiography.

5.2.6 Additional biochemical methods

5.2.6.1 Chromatin immunoprecipitation (ChIP)

Chromatin immunoprecipitation was performed in three independent experiments for each protein mainly as described (Hecht and Grunstein, 1999). Cells from 45 ml liquid culture were crosslinked with formaldehyde as described (see section 5.2.1.4), washed with 500 µl cold ChIP lysis buffer [50 mM HEPES pH 7.5, 140 mM NaCl, 5 mM EDTA pH 8.0, 5 mM EGTA pH 8.0, 1% (v/v) Triton X-100, 0.1% (w/v) DOC] and resuspended in an equal volume of cold ChIP lysis buffer. An equal volume of cold glass beads (∅ 0.75-1 mm) was added to the suspension and the tube was vigorously shaken on an IKA-Vibrax VXR for 3x 15 min at 2200 rpm and 4°C. The tube was pierced on the bottom and on the lid with a hot needle and placed in a 15 ml tube to remove the glass beads from the cell extract by centrifugation for 2 min at 2000 rpm and 4°C. Cold ChIP lysis buffer was added to a final volume of 1 ml and the DNA in the suspension was sonicated using a Sonifier 250 to obtain an average DNA fragment size of 500-1000 bp. Cell debris were removed by centrifugation for 20 min at 13000 rpm and 4°C. The chromatin extracts were split into three aliquots. 40 µl of each aliquot served as an input control. 200 µl of each aliquot were incubated either with 3 µg of a monoclonal α-HA antibody (3F10) and 50 µl equilibrated Protein G Sepharose or with 50 µl equilibrated IgG Sepharose for 2 h at 4°C to enrich either HA_3-tagged proteins or Prot.A-tagged proteins. After immunoprecipitation, the beads were washed 3x with cold ChIP lysis buffer, 2x with cold ChIP washing buffer I [50 mM HEPES pH 7.5, 500 mM NaCl, 2 mM EDTA pH 8.0, 1% (v/v) Triton X-100, 0.1% (w/v) DOC], and 2x with cold ChIP washing buffer II [10 mM Tris-HCl pH 8.0, 250 mM LiCl, 2 mM EDTA pH 8.0, 0.5% (v/v) Nonidet P40, 0.5% (w/v) DOC] followed by a final washing step with TE buffer [10 mM Tris-HCl pH 8.0, 1 mM EDTA pH 8.0]. 250 µl IRN buffer [50 mM Tris-HCl pH 8.0, 20 mM EDTA pH 8.0, 500 mM NaCl] were added to the input (IN) and to the beads (IP) samples. RNA in the samples was digested with 2 µl RNase A (20 mg/ml) for 1 h at 37°C. Afterwards, 0.5% (w/v) SDS and 2 µl Proteinase K (20 mg/ml) were added to the samples followed by incubation for 1 h at 56°C. The formaldehyde crosslink was reversed overnight by incubation at 65°C. DNA was isolated by phenol-chloroform extraction as described (see section 5.2.3.1), precipitated with ethanol as described (see section 5.2.3.2) and resuspended in 50 µl TE buffer.

DNA amounts present in IN and IP samples were determined by quantitative real-time PCR as described (see section 5.2.3.8). Primer pairs for amplification used in this work are listed in section 5.1.4. IN DNA was diluted 1:1000 and IP DNA was diluted 1:200 prior to analysis. Retention of specific DNA fragments was calculated as the fraction of the total input DNA. The mean values and error bars were derived from three independent ChIP experiments analyzed in triplicate quantitative real-time PCR reactions.

5.2.6.2 Chromatin endogenous cleavage (ChEC)

Chromatin endogenous cleavage was performed mainly as described (Schmid et al., 2004). Cells from 45 ml liquid culture were crosslinked with formaldehyde as described (see section 5.2.1.4), washed 3x with 500 µl cold ChEC buffer A (+ 1x PIs), and resuspended in 350 µl cold ChEC buffer A (+ 1x PIs). An equal volume of cold glass beads (∅ 0.75-1 mm) was added to the suspension and the tube was vigorously shaken on an IKA-Vibrax VXR for 10 min at 2200 rpm and 4°C. The tube was pierced on the bottom and on the lid with a hot needle and placed in a 15 ml tube to remove the glass beads from the cell extract by centrifugation for 2 min at 2000 rpm and 4°C. The cell lysate was transferred to a 1.5 ml tube and centrifuged again for 2 min at 13000 rpm and 4°C. The supernatant was discarded. The lysate pellet containing intact nuclei was washed with 500 µl cold ChEC buffer A (+ 1x PIs) and resuspended in 600 µl cold ChEC buffer Ag (+ 1x PIs). The sample was incubated in a thermomixer at 700 rpm and 30°C. Two aliquots (0 min, untreated) of the well-mixed sample were taken (100 µl). MNase digestion was started by adding 8 µl 100 mM (w/v) $CaCl_2$. Aliquots were taken at 10 min, 20 min, and 60 min (100 µl). The digestion reaction was immediately stopped by pipetting the aliquot into 100 µl IRN buffer [50 mM Tris-HCl pH 8.0, 20 mM EDTA pH 8.0, 500 mM NaCl]. Before taking an aliquot, the sample should be mixed at higher rotation rates since nuclei sediment. Stopped aliquots could be kept at RT. When the time course was complete, 100 µl were added to the 0 min aliquots. RNA in the aliquots was digested with 2 µl RNase A (20 mg/ml) for 1 h at 37°C. Afterwards, 0.5% (w/v) SDS and 2 µl Proteinase K (20 mg/ml) were added to the aliquots followed by incubation for 1 h at 56°C. The formaldehyde crosslink was reversed overnight by incubation at 65°C. DNA was isolated by phenol-chloroform extraction as described (see section 5.2.3.1), precipitated with ethanol as described (see section 5.2.3.2), and resuspended in 50 µl TE buffer.

20 µl of the DNA sample were digested with the restriction endonuclease XcmI in a 27 µl reaction overnight at 37°C. The restriction endonuclease XcmI can cleave cryptic recognition sites if present in excess, so the amount of this enzyme should be controlled well (0.5 µl of XcmI per aliquot). Total DNA of the digestion reaction was separated by native agarose gel electrophoresis as described (see section 5.2.3.4), transferred to a membrane as described (see section 5.2.3.5), and analyzed as described (see section 5.2.3.6).

5.2.6.3 Co-immunoprecipitation (CoIP)

In one case, cells were grown in SCR medium depleted of leucine at 30°C before galactose was added and cells were shifted for 8 h to 24°C. Cells from 45 ml liquid culture were crosslinked with formaldehyde as described (see section 5.2.1.4), washed with 500 µl cold ChIP lysis buffer [50 mM HEPES pH 7.5, 140 mM NaCl, 5 mM EDTA pH 8.0, 5 mM EGTA pH 8.0, 1% (v/v) Triton X-100, 0.1% (w/v) DOC], and resuspended in an equal volume of cold ChIP lysis buffer. An equal volume of cold glass beads (∅ 0.75-1 mm) was added to the suspension and the tube was vigorously shaken on an IKA-Vibrax VXR for 3x 15 min at 2200 rpm and 4°C. The tube was pierced on the bottom and on the lid with a hot needle and placed in a 15 ml tube to remove the glass beads from the cell extract by centrifugation for 2 min at 2000 rpm and 4°C. Cold ChIP lysis buffer was added to a final volume of 1 ml and DNA in the suspension was sonicated using a Sonifier 250. Cell debris were removed by centrifugation for 20 min at 13000 rpm and 4°C. Same amounts of the chromatin extract were either precipitated (21 µl) using methanol-chloroform precipitation as described (see section 5.2.5.4) and served as an input control or incubated (700 µl) with 3 µg of a monoclonal α-HA antibody (3F10) and 100 µl equilibrated Protein G Sepharose for 2 h at 4°C to enrich HA_3-tagged proteins. After immunoprecipitation, the beads were washed 3x with cold ChIP lysis buffer, 3x with cold ChIP washing buffer I [50 mM HEPES pH 7.5, 500 mM NaCl, 2 mM EDTA pH 8.0, 1% (v/v) Triton X-100, 0.1% (w/v) DOC], and 3x with cold ChIP washing buffer II [10 mM Tris-HCl pH 8.0, 250 mM LiCl, 2 mM EDTA pH 8.0, 0.5% (v/v) Nonidet P40, 0.5% (w/v) DOC] followed by a final washing step with TE buffer [10 mM Tris-HCl pH 8.0, 1 mM EDTA pH 8.0]. 30 µl 4x SDS sample buffer were added to the beads (IP) and proteins were eluted by incubating the sample for 10 min at 99°C. 30 µl 4x SDS sample buffer were added to the protein pellet of the input (IN) fraction and both the IN and IP samples were incubated for 30 min at 99°C to reverse the formaldehyde crosslink. Appropriate amounts of the samples were analyzed as described (see sections 5.2.5.6, 5.2.5.7, and 5.2.5.8).

In the other case, cells were grown in YPD medium at 30°C. Cells from 50 ml liquid culture were washed with 1 ml cold H_2O, transferred to a 1.5 ml tube, and resuspended in an equal volume of cold lysis buffer [20 mM HEPES pH 7.6, 150 mM KAc, 5 mM MgOAc, 2 mM benzamidine, 1 mM PMSF]. An equal volume of cold glass beads (∅ 0.75-1 mm) was added to the suspension and the tube was vigorously shaken in a FastPrep Instrument for 2x 40 sec at 5.5 m/s and 4°C. The tube was pierced on the bottom and on the lid with a hot needle and placed in a 15 ml tube to remove the glass beads from the cell extract by centrifugation for 2 min at 2000 rpm and 4°C. The cell extract was transferred to a 1.5 ml tube and centrifuged again for 15 min at 13000 rpm and 4°C to remove cell debris. Protein concentration of the cleared supernatant (WCE) was determined using the Bradford assay (Bradford, 1976). Same amounts of WCE (5 mg) were supplemented with 0.5% (v/v) Triton X-100 and 0.1% (v/v) Tween 20. Afterwards, cold binding buffer [20 mM HEPES pH 7.6, 150 mM KAc, 5 mM MgOAc, 2 mM benzamidine, 1 mM PMSF, 0.5% (v/v) Triton X-100, 0.1% (v/v) Tween 20] was added to a final volume of 1 ml. 5 µl of the suspension served as an input control, whereas the residual suspension was incubated with 100 µl equilibrated IgG magnetic beads for 2 h at 4°C to enrich Prot.A-tagged proteins. After immunoprecipitation, the beads were washed 3x with cold binding buffer followed by a final washing step with TE buffer [10 mM Tris-HCl pH 8.0, 1 mM EDTA pH 8.0]. 5 µl 4x SDS sample buffer were added to the input (IN) fraction and 50 µl 4x SDS sample buffer were added to the beads (IP). Both the IN and IP samples were incubated for 5 min at 99°C and appropriate amounts of the samples were analyzed as described (see sections 5.2.5.6, 5.2.5.7, and 5.2.5.8).

5.2.6.4 Calmodulin affinity precipitation

Cells were grown in YPD medium at 30°C. Whole cell extract from 10 l liquid culture was prepared as described (see section 5.2.5.1.2). Same amounts of WCE (1.2 g) were mixed with an equal volume of dilution buffer [20% (v/v) glycerol, 150 mM HEPES pH 7.6, 10 mM $MgCl_2$, 400 mM $(NH_4)_2SO_4$, 15 mM β-mercaptoethanol, 2 mM benzamidine, 1 mM PMSF, 600 mM NaCl, 2 mM imidazole, 4 mM $CaCl_2$, 0.2% (v/v) Nonidet P40] and incubated with 500 µl equilibrated Calmodulin Affinity Resin for 2 h at 4°C to enrich TAP-tagged proteins. After affinity precipitation, the beads were washed with cold binding buffer [20% (v/v) glycerol, 150 mM HEPES pH 7.6, 10 mM $MgCl_2$, 400 mM $(NH_4)_2SO_4$, 10 mM β-mercaptoethanol, 2 mM benzamidine, 1 mM PMSF, 300 mM NaCl, 1 mM imidazole, 2 mM $CaCl_2$, 0.1% (v/v) Nonidet P40] and with cold washing buffer [100 mM HEPES pH 7.6, 20 mM NaCl, 10 mM β-mercaptoethanol, 1 mM imidazole, 0.1% (v/v) Nonidet P40]. Proteins were eluted with cold elution buffer [100 mM HEPES pH 7.6, 20 mM NaCl, 10 mM β-mercaptoethanol, 1 mM imidazole, 0.1% (v/v) Nonidet P40, 5 mM EGTA pH 8.0].

MATERIAL AND METHODS

5.2.6.5 Anion exchange chromatography

Elution fractions from the calmodulin affinity precipitation were pooled and loaded on a Mono Q PC 1.6/5 column using a SMART System. The column was equilibrated with buffer A [5% (v/v) glycerol, 20 mM HEPES pH 7.6, 1 mM DTT, 0.5% (v/v) Nonidet P40, 0.1% (v/v) Tween 20] and eluted with a linear gradient from 0 M to 1 M NaCl using buffer B [1 M NaCl, 5% (v/v) glycerol, 20 mM HEPES pH 7.6, 1 mM DTT, 0.5% (v/v) Nonidet P40, 0.1% (v/v) Tween 20].

5.2.6.6 Gel filtration chromatography

Elution fractions from the anion exchange chromatography were pooled and a fraction was loaded on a Superose 12 PC 3.2/30 column using a SMART System. The column was equilibrated with buffer C [200 mM NaCl, 5% (v/v) glycerol, 20 mM HEPES pH 7.6, 1 mM DTT, 0.05% (v/v) Nonidet P40, 0.05% (v/v) Tween 20].

5.2.6.7 Affinity precipitation of ubiquitylated proteins

In one case, cells were grown in YPD or YPG medium, respectively, at 30°C and further cultivated either in the absence or in the presence of rapamycin. Whole cell extract from 50 ml was prepared as described (see section 5.2.5.1.1). Same amounts of WCE (6 mg) were supplemented with 750 mM KAc, 0.5% (v/v) Nonidet P40, and 0.05% (v/v) Triton X-100 and cold high-salt extraction buffer [150 mM HEPES pH 7.6, 400 mM $(NH_4)_2SO_4$, 10 mM $MgCl_2$, 20% (v/v) glycerol, 5 mM β-mercaptoethanol, 2 mM benzamidine, 1 mM PMSF] was added to a final volume of 1.2 ml. 12 µl of the suspension served as an input control, whereas the residual suspension was incubated either with recombinant GST or with recombinant GST-Dsk2p bound to 50 µl equilibrated Glutathione Sepharose overnight at 4°C to enrich (poly)ubiquitylated proteins. After affinity precipitation, the beads were washed 2x with cold washing buffer [750 mM KAc, 0.5% (v/v) Nonidet P40] and 2x with cold 1x PBS. 50 µl 4x SDS sample buffer were added to the beads (IP) and proteins were eluted by incubating the sample for 5 min at 99°C. Alternatively, proteins were eluted with cold elution buffer [50 mM Tris-HCl pH 8.0, 150 mM NaCl, 10 mM reduced glutathione]. 5 µl 4x SDS sample buffer were added to the input (IN) fraction, to the corresponding flow through (FT) fraction, and to the corresponding wash fractions. Appropriate amounts of the samples were analyzed as described (see sections 5.2.5.6, 5.2.5.7, and 5.2.5.8).

In the other case, cells were grown in YPD at 24°C before cells were shifted for 2 h to 37°C and further cultivated either in the absence or in the presence of rapamycin. Whole cell extract from 200 ml liquid culture was prepared as described (see section 5.2.5.1.1). Same amounts of WCE (20 mg) were supplemented with 750 mM KAc, 0.5% (v/v) Nonidet P40, and 0.05% (v/v) Triton X-100. Afterwards, cold high-salt extraction buffer [150 mM HEPES pH 7.6, 400 mM $(NH_4)_2SO_4$, 10 mM $MgCl_2$, 20% (v/v) glycerol, 5 mM β-mercaptoethanol, 2 mM benzamidine, 1 mM PMSF] was added to a final volume of 1.2 ml. The suspension was incubated either with recombinant GST or with recombinant GST-Dsk2p bound to 50 µl equilibrated Glutathione Sepharose overnight at 4°C to enrich (poly)ubiquitylated proteins. After affinity precipitation, the beads were washed 2x with cold washing buffer [750 mM KAc, 0.5% (v/v) Nonidet P40] and 2x with cold 1x PBS. 30 µl 4x SDS sample buffer were added to the beads (IP) and proteins were eluted by incubating the sample for 5 min at 99°C. Appropriate amounts of the samples were analyzed as described (see sections 5.2.5.6, 5.2.5.7, and 5.2.5.8).

5.2.6.8 Live cell imaging

Treated cells were harvested by centrifugation in 1.5 ml tube for 10 sec at 6000 rpm and RT before the cell pellet was washed either with SCD (cells grown in media based on glucose) or SCG (cells grown in media based on galactose). Cells were immediately analyzed by fluorescence microscopy.

Images were captured with an AxioCam MR CCD camera mounted on an Axiovert 200M microscope and processed with Axiovision V 4.7.1.0 and Adobe Photoshop CS v.8.0.1.

MATERIAL AND METHODS

6 REFERENCES

Abovich, N., Gritz, L., Tung, L., and Rosbash, M. (1985). Effect of RP51 gene dosage alterations on ribosome synthesis in Saccharomyces cerevisiae. Mol. Cell. Biol *5*, 3429-3435.

Aprikian, P., Moorefield, B., and Reeder, R. H. (2001). New model for the yeast RNA polymerase I transcription cycle. Mol. Cell. Biol *21*, 4847-4855.

Barbet, N. C., Schneider, U., Helliwell, S. B., Stansfield, I., Tuite, M. F., and Hall, M. N. (1996). TOR controls translation initiation and early G1 progression in yeast. Mol. Biol. Cell *7*, 25-42.

Bateman, E., and Paule, M. R. (1986). Regulation of eukaryotic ribosomal RNA transcription by RNA polymerase modification. Cell *47*, 445-450.

Beckouet, F., Labarre-Mariotte, S., Albert, B., Imazawa, Y., Werner, M., Gadal, O., Nogi, Y., and Thuriaux, P. (2008). Two RNA polymerase I subunits control the binding and release of Rrn3 during transcription. Mol. Cell. Biol *28*, 1596-1605.

van Beekvelt, C. A., de Graaff-Vincent, M., Faber, A. W., van't Riet, J., Venema, J., and Raué, H. A. (2001). All three functional domains of the large ribosomal subunit protein L25 are required for both early and late pre-rRNA processing steps in Saccharomyces cerevisiae. Nucleic Acids Res *29*, 5001-5008.

van Beekvelt, C. A., Kooi, E. A., de Graaff-Vincent, M., Riet, J., Venema, J., and Raué, H. A. (2000). Domain III of Saccharomyces cerevisiae 25 S ribosomal RNA: its role in binding of ribosomal protein L25 and 60 S subunit formation. J. Mol. Biol *296*, 7-17.

Bell, G. I., Valenzuela, P., and Rutter, W. J. (1977). Phosphorylation of yeast DNA-dependent RNA polymerases in vivo and in vitro. Isolation of enzymes and identification of phosphorylated subunits. J. Biol. Chem *252*, 3082-3091.

Bell, G. I., Valenzuela, P., and Rutter, W. J. (1976). Phosphorylation of yeast RNA polymerases. Nature *261*, 429-431.

Bensadoun, A., and Weinstein, D. (1976). Assay of proteins in the presence of interfering materials. Anal. Biochem *70*, 241-250.

Beretta, L., Gingras, A. C., Svitkin, Y. V., Hall, M. N., and Sonenberg, N. (1996). Rapamycin blocks the phosphorylation of 4E-BP1 and inhibits cap-dependent initiation of translation. EMBO J *15*, 658-664.

Berger, A. B., Decourty, L., Badis, G., Nehrbass, U., Jacquier, A., and Gadal, O. (2007). Hmo1 is required for TOR-dependent regulation of ribosomal protein gene transcription. Mol. Cell. Biol *27*, 8015-8026.

Berset, C., Trachsel, H., and Altmann, M. (1998). The TOR (target of rapamycin) signal transduction pathway regulates the stability of translation initiation factor eIF4G in the yeast

REFERENCES

Saccharomyces cerevisiae. Proc. Natl. Acad. Sci. U.S.A 95, 4264-4269.

Bier, M., Fath, S., and Tschochner, H. (2004). The composition of the RNA polymerase I transcription machinery switches from initiation to elongation mode. FEBS Lett 564, 41-46.

Birch, J. L., Tan, B. C., Panov, K. I., Panova, T. B., Andersen, J. S., Owen-Hughes, T. A., Russell, J., Lee, S., and Zomerdijk, J. C. B. M. (2009). FACT facilitates chromatin transcription by RNA polymerases I and III. EMBO J 28, 854-865.

Bodem, J., Dobreva, G., Hoffmann-Rohrer, U., Iben, S., Zentgraf, H., Delius, H., Vingron, M., and Grummt, I. (2000). TIF-IA, the factor mediating growth-dependent control of ribosomal RNA synthesis, is the mammalian homolog of yeast Rrn3p. EMBO Rep 1, 171-175.

Bouchoux, C., Hautbergue, G., Grenetier, S., Carles, C., Riva, M., and Goguel, V. (2004). CTD kinase I is involved in RNA polymerase I transcription. Nucleic Acids Res 32, 5851-5860.

Bradford, M. M. (1976). A rapid and sensitive method for the quantitation of microgram quantities of protein utilizing the principle of protein-dye binding. Anal. Biochem 72, 248-254.

Braglia, P., Kawauchi, J., and Proudfoot, N. J. (2010). Co-transcriptional RNA cleavage provides a failsafe termination mechanism for yeast RNA polymerase I. Nucleic Acids Res. Available at: http://www.ncbi.nlm.nih.gov/pubmed/20972219 [Accessed November 10, 2010].

Bréant, B., Buhler, J. M., Sentenac, A., and Fromageot, P. (1983). On the phosphorylation of yeast RNA polymerases A and B. Eur. J. Biochem 130, 247-251.

Brun, R. P., Ryan, K., and Sollner-Webb, B. (1994). Factor C*, the specific initiation component of the mouse RNA polymerase I holoenzyme, is inactivated early in the transcription process. Mol. Cell. Biol 14, 5010-5021.

Buhler, J. M., Huet, J., Davies, K. E., Sentenac, A., and Fromageot, P. (1980). Immunological studies of yeast nuclear RNA polymerases at the subunit level. J. Biol. Chem 255, 9949-9954.

Buhler, J. M., Iborra, F., Sentenac, A., and Fromageot, P. (1976). The presence of phosphorylated subunits in yeast RNA polymerases A and B. FEBS Lett 72, 37-41.

Buttgereit, D., Pflugfelder, G., and Grummt, I. (1985). Growth-dependent regulation of rRNA synthesis is mediated by a transcription initiation factor (TIF-IA). Nucleic Acids Res 13, 8165-8180.

Cadwell, C., Yoon, H. J., Zebarjadian, Y., and Carbon, J. (1997). The yeast nucleolar protein Cbf5p is involved in rRNA biosynthesis and interacts genetically with the RNA polymerase I transcription factor RRN3. Mol. Cell. Biol 17, 6175-6183.

Cardenas, M. E., Cutler, N. S., Lorenz, M. C., Di Como, C. J., and Heitman, J. (1999). The TOR signaling cascade regulates gene expression in response to nutrients. Genes Dev 13, 3271-3279.

REFERENCES

Carles, C., and Riva, M. (1998). Yeast RNA polymerase I subunits and genes. In Transcription of ribosomal RNA genes by eukaryotic RNA polymerase I (Paule, M. R. Ed.), pp. 9-38. Landes Bioscience, Austin.

Carles, C., Treich, I., Bouet, F., Riva, M., and Sentenac, A. (1991). Two additional common subunits, ABC10 alpha and ABC10 beta, are shared by yeast RNA polymerases. J. Biol. Chem 266, 24092-24096.

Cavanaugh, A. H., Evans, A., and Rothblum, L. I. (2008). Mammalian Rrn3 is required for the formation of a transcription competent preinitiation complex containing RNA polymerase I. Gene Expr 14, 131-147.

Cavanaugh, A. H., Hirschler-Laszkiewicz, I., Hu, Q., Dundr, M., Smink, T., Misteli, T., and Rothblum, L. I. (2002). Rrn3 phosphorylation is a regulatory checkpoint for ribosome biogenesis. J. Biol. Chem 277, 27423-27432.

Choe, S. Y., Schultz, M. C., and Reeder, R. H. (1992). In vitro definition of the yeast RNA polymerase I promoter. Nucleic Acids Res 20, 279-285.

Cioci, F., Vu, L., Eliason, K., Oakes, M., Siddiqi, I. N., and Nomura, M. (2003). Silencing in yeast rDNA chromatin: reciprocal relationship in gene expression between RNA polymerase I and II. Mol. Cell 12, 135-145.

Claypool, J. A., French, S. L., Johzuka, K., Eliason, K., Vu, L., Dodd, J. A., Beyer, A. L., and Nomura, M. (2004). Tor pathway regulates Rrn3p-dependent recruitment of yeast RNA polymerase I to the promoter but does not participate in alteration of the number of active genes. Mol. Biol. Cell 15, 946-956.

Clos, J., Buttgereit, D., and Grummt, I. (1986). A purified transcription factor (TIF-IB) binds to essential sequences of the mouse rDNA promoter. Proc. Natl. Acad. Sci. U.S.A 83, 604-608.

Conconi, A., Widmer, R. M., Koller, T., and Sogo, J. M. (1989). Two different chromatin structures coexist in ribosomal RNA genes throughout the cell cycle. Cell 57, 753-761.

Dammann, R., Lucchini, R., Koller, T., and Sogo, J. M. (1993). Chromatin structures and transcription of rDNA in yeast Saccharomyces cerevisiae. Nucleic Acids Res 21, 2331-2338.

De Virgilio, C., and Loewith, R. (2006a). Cell growth control: little eukaryotes make big contributions. Oncogene 25, 6392-6415.

De Virgilio, C., and Loewith, R. (2006b). The TOR signalling network from yeast to man. Int. J. Biochem. Cell Biol 38, 1476-1481.

Deutschbauer, A. M., Jaramillo, D. F., Proctor, M., Kumm, J., Hillenmeyer, M. E., Davis, R. W., Nislow, C., and Giaever, G. (2005). Mechanisms of haploinsufficiency revealed by genome-wide profiling in yeast. Genetics 169, 1915-1925.

Di Como, C. J., and Arndt, K. T. (1996). Nutrients, via the Tor proteins, stimulate the association of Tap42 with type 2A phosphatases. Genes Dev 10, 1904-1916.

REFERENCES

El Hage, A., Koper, M., Kufel, J., and Tollervey, D. (2008). Efficient termination of transcription by RNA polymerase I requires the 5' exonuclease Rat1 in yeast. Genes Dev 22, 1069-1081.

Elion, E. A., and Warner, J. R. (1986). An RNA polymerase I enhancer in Saccharomyces cerevisiae. Mol. Cell. Biol 6, 2089-2097.

Elion, E. A., and Warner, J. R. (1984). The major promoter element of rRNA transcription in yeast lies 2 kb upstream. Cell 39, 663-673.

Farrar, J. E., Nater, M., Caywood, E., McDevitt, M. A., Kowalski, J., Takemoto, C. M., Talbot, C. C., Meltzer, P., Esposito, D., Beggs, A. H., et al. (2008). Abnormalities of the large ribosomal subunit protein, Rpl35a, in Diamond-Blackfan anemia. Blood 112, 1582-1592.

Fath, S., Milkereit, P., Peyroche, G., Riva, M., Carles, C., and Tschochner, H. (2001). Differential roles of phosphorylation in the formation of transcriptional active RNA polymerase I. Proc. Natl. Acad. Sci. U.S.A 98, 14334-14339.

Fath, S. (2002). Untersuchungen zur Regulation der Ribosomen-Biogenese in Saccharomyces cerevisiae.

Fath, S., Kobor, M. S., Philippi, A., Greenblatt, J., and Tschochner, H. (2004). Dephosphorylation of RNA polymerase I by Fcp1p is required for efficient rRNA synthesis. J. Biol. Chem 279, 25251-25259.

Fatica, A., and Tollervey, D. (2002). Making ribosomes. Curr. Opin. Cell Biol 14, 313-318.

Fátyol, K., and Grummt, I. (2008). Proteasomal ATPases are associated with rDNA: the ubiquitin proteasome system plays a direct role in RNA polymerase I transcription. Biochim. Biophys. Acta 1779, 850-859.

Favier, D., and Gonda, T. J. (1994). Detection of proteins that bind to the leucine zipper motif of c-Myb. Oncogene 9, 305-311.

Ferreira-Cerca, S., Pöll, G., Gleizes, P., Tschochner, H., and Milkereit, P. (2005). Roles of eukaryotic ribosomal proteins in maturation and transport of pre-18S rRNA and ribosome function. Mol. Cell 20, 263-275.

Ferreira-Cerca, S., Pöll, G., Kühn, H., Neueder, A., Jakob, S., Tschochner, H., and Milkereit, P. (2007). Analysis of the in vivo assembly pathway of eukaryotic 40S ribosomal proteins. Mol. Cell 28, 446-457.

Fromont-Racine, M., Senger, B., Saveanu, C., and Fasiolo, F. (2003). Ribosome assembly in eukaryotes. Gene 313, 17-42.

Funakoshi, M., Sasaki, T., Nishimoto, T., and Kobayashi, H. (2002). Budding yeast Dsk2p is a polyubiquitin-binding protein that can interact with the proteasome. Proc. Natl. Acad. Sci. U.S.A 99, 745-750.

Gadal, O., Labarre, S., Boschiero, C., and Thuriaux, P. (2002). Hmo1, an HMG-box protein, belongs

REFERENCES

to the yeast ribosomal DNA transcription system. EMBO J 21, 5498-5507.

Gallagher, J. E. G., Dunbar, D. A., Granneman, S., Mitchell, B. M., Osheim, Y., Beyer, A. L., and Baserga, S. J. (2004). RNA polymerase I transcription and pre-rRNA processing are linked by specific SSU processome components. Genes Dev 18, 2506-2517.

Garí, E., Piedrafita, L., Aldea, M., and Herrero, E. (1997). A set of vectors with a tetracycline-regulatable promoter system for modulated gene expression in Saccharomyces cerevisiae. Yeast 13, 837-848.

Geiger, S. R., Kuhn, C. D., Leidig, C., Renkawitz, J., and Cramer, P. (2008). Crystallization of RNA polymerase I subcomplex A14/A43 by iterative prediction, probing and removal of flexible regions. Acta Crystallogr. Sect. F Struct. Biol. Cryst. Commun 64, 413-418.

Geiger, S. R., Lorenzen, K., Schreieck, A., Hanecker, P., Kostrewa, D., Heck, A. J. R., and Cramer, P. (2010). RNA polymerase I contains a TFIIF-related DNA-binding subcomplex. Mol. Cell 39, 583-594.

Gerbasi, V. R., Weaver, C. M., Hill, S., Friedman, D. B., and Link, A. J. (2004). Yeast Asc1p and mammalian RACK1 are functionally orthologous core 40S ribosomal proteins that repress gene expression. Mol. Cell. Biol 24, 8276-8287.

Gerber, J., Reiter, A., Steinbauer, R., Jakob, S., Kuhn, C., Cramer, P., Griesenbeck, J., Milkereit, P., and Tschochner, H. (2008). Site specific phosphorylation of yeast RNA polymerase I. Nucleic Acids Res 36, 793-802.

Gerlinger, U. M., Gückel, R., Hoffmann, M., Wolf, D. H., and Hilt, W. (1997). Yeast cycloheximide-resistant crl mutants are proteasome mutants defective in protein degradation. Mol. Biol. Cell 8, 2487-2499.

Gorenstein, C., and Warner, J. R. (1977). Synthesis and turnover of ribosomal proteins in the absence of 60S subunit assembly in Saccharomyces cerevisiae. Mol. Gen. Genet 157, 327-332.

Granneman, S., and Baserga, S. J. (2005). Crosstalk in gene expression: coupling and co-regulation of rDNA transcription, pre-ribosome assembly and pre-rRNA processing. Curr. Opin. Cell Biol 17, 281-286.

Grummt, I., Smith, V. A., and Grummt, F. (1976). Amino acid starvation affects the initiation frequency of nucleolar RNA polymerase. Cell 7, 439-445.

Grummt, I. (2003). Life on a planet of its own: regulation of RNA polymerase I transcription in the nucleolus. Genes Dev 17, 1691-1702.

Hall, D. B., Wade, J. T., and Struhl, K. (2006). An HMG protein, Hmo1, associates with promoters of many ribosomal protein genes and throughout the rRNA gene locus in Saccharomyces cerevisiae. Mol. Cell. Biol 26, 3672-3679.

Hannan, K. M., Brandenburger, Y., Jenkins, A., Sharkey, K., Cavanaugh, A., Rothblum, L., Moss, T.,

REFERENCES

Poortinga, G., McArthur, G. A., Pearson, R. B., et al. (2003). mTOR-dependent regulation of ribosomal gene transcription requires S6K1 and is mediated by phosphorylation of the carboxy-terminal activation domain of the nucleolar transcription factor UBF. Mol. Cell. Biol 23, 8862-8877.

Hartwell, L. H., McLaughlin, C. S., and Warner, J. R. (1970). Identification of ten genes that control ribosome formation in yeast. Mol. Gen. Genet 109, 42-56.

Hay, N., and Sonenberg, N. (2004). Upstream and downstream of mTOR. Genes Dev 18, 1926-1945.

Hecht, A., and Grunstein, M. (1999). Mapping DNA interaction sites of chromosomal proteins using immunoprecipitation and polymerase chain reaction. Meth. Enzymol 304, 399-414.

Heitman, J., Movva, N. R., and Hall, M. N. (1991). Targets for cell cycle arrest by the immunosuppressant rapamycin in yeast. Science 253, 905-909.

Henderson, A. S., Warburton, D., and Atwood, K. C. (1972). Location of ribosomal DNA in the human chromosome complement. Proc. Natl. Acad. Sci. U.S.A 69, 3394-3398.

Henras, A. K., Soudet, J., Gérus, M., Lebaron, S., Caizergues-Ferrer, M., Mougin, A., and Henry, Y. (2008). The post-transcriptional steps of eukaryotic ribosome biogenesis. Cell. Mol. Life Sci 65, 2334-2359.

Hirschler-Laszkiewicz, I., Cavanaugh, A. H., Mirza, A., Lun, M., Hu, Q., Smink, T., and Rothblum, L. I. (2003). Rrn3 becomes inactivated in the process of ribosomal DNA transcription. J. Biol. Chem 278, 18953-18959.

Honma, Y., Kitamura, A., Shioda, R., Maruyama, H., Ozaki, K., Oda, Y., Mini, T., Jenö, P., Maki, Y., Yonezawa, K., et al. (2006). TOR regulates late steps of ribosome maturation in the nucleoplasm via Nog1 in response to nutrients. EMBO J 25, 3832-3842.

Hoppe, S., Bierhoff, H., Cado, I., Weber, A., Tiebe, M., Grummt, I., and Voit, R. (2009). AMP-activated protein kinase adapts rRNA synthesis to cellular energy supply. Proc. Natl. Acad. Sci. U.S.A 106, 17781-17786.

Huang, J., Zhu, H., Haggarty, S. J., Spring, D. R., Hwang, H., Jin, F., Snyder, M., and Schreiber, S. L. (2004). Finding new components of the target of rapamycin (TOR) signaling network through chemical genetics and proteome chips. Proc. Natl. Acad. Sci. U.S.A 101, 16594-16599.

Huber, A., Bodenmiller, B., Uotila, A., Stahl, M., Wanka, S., Gerrits, B., Aebersold, R., and Loewith, R. (2009). Characterization of the rapamycin-sensitive phosphoproteome reveals that Sch9 is a central coordinator of protein synthesis. Genes Dev 23, 1929-1943.

Iben, S., Tschochner, H., Bier, M., Hoogstraten, D., Hozák, P., Egly, J. M., and Grummt, I. (2002). TFIIH plays an essential role in RNA polymerase I transcription. Cell 109, 297-306.

Janke, C., Magiera, M. M., Rathfelder, N., Taxis, C., Reber, S., Maekawa, H., Moreno-Borchart, A.,

REFERENCES

Doenges, G., Schwob, E., Schiebel, E., et al. (2004). A versatile toolbox for PCR-based tagging of yeast genes: new fluorescent proteins, more markers and promoter substitution cassettes. Yeast *21*, 947-962.

Jansa, P., and Grummt, I. (1999). Mechanism of transcription termination: PTRF interacts with the largest subunit of RNA polymerase I and dissociates paused transcription complexes from yeast and mouse. Mol. Gen. Genet *262*, 508-514.

Jantzen, H. M., Admon, A., Bell, S. P., and Tjian, R. (1990). Nucleolar transcription factor hUBF contains a DNA-binding motif with homology to HMG proteins. Nature *344*, 830-836.

Jiang, Y., and Broach, J. R. (1999). Tor proteins and protein phosphatase 2A reciprocally regulate Tap42 in controlling cell growth in yeast. EMBO J *18*, 2782-2792.

Jorgensen, P., Nishikawa, J. L., Breitkreutz, B., and Tyers, M. (2002). Systematic identification of pathways that couple cell growth and division in yeast. Science *297*, 395-400.

Jorgensen, P., Rupes, I., Sharom, J. R., Schneper, L., Broach, J. R., and Tyers, M. (2004). A dynamic transcriptional network communicates growth potential to ribosome synthesis and critical cell size. Genes Dev *18*, 2491-2505.

Ju, Q., and Warner, J. R. (1994). Ribosome synthesis during the growth cycle of Saccharomyces cerevisiae. Yeast *10*, 151-157.

Kaeberlein, M., Powers, R. W., Steffen, K. K., Westman, E. A., Hu, D., Dang, N., Kerr, E. O., Kirkland, K. T., Fields, S., and Kennedy, B. K. (2005). Regulation of yeast replicative life span by TOR and Sch9 in response to nutrients. Science *310*, 1193-1196.

Kawauchi, J., Mischo, H., Braglia, P., Rondon, A., and Proudfoot, N. J. (2008). Budding yeast RNA polymerases I and II employ parallel mechanisms of transcriptional termination. Genes Dev *22*, 1082-1092.

Keener, J., Dodd, J. A., Lalo, D., and Nomura, M. (1997). Histones H3 and H4 are components of upstream activation factor required for the high-level transcription of yeast rDNA by RNA polymerase I. Proc. Natl. Acad. Sci. U.S.A *94*, 13458-13462.

Keough, R. A., Macmillan, E. M., Lutwyche, J. K., Gardner, J. M., Tavner, F. J., Jans, D. A., Henderson, B. R., and Gonda, T. J. (2003). Myb-binding protein 1a is a nucleocytoplasmic shuttling protein that utilizes CRM1-dependent and independent nuclear export pathways. Exp. Cell Res *289*, 108-123.

Keys, D. A., Lee, B. S., Dodd, J. A., Nguyen, T. T., Vu, L., Fantino, E., Burson, L. M., Nogi, Y., and Nomura, M. (1996). Multiprotein transcription factor UAF interacts with the upstream element of the yeast RNA polymerase I promoter and forms a stable preinitiation complex. Genes Dev *10*, 887-903.

Keys, D. A., Vu, L., Steffan, J. S., Dodd, J. A., Yamamoto, R. T., Nogi, Y., and Nomura, M. (1994). RRN6 and RRN7 encode subunits of a multiprotein complex essential for the initiation of rDNA transcription by RNA polymerase I in Saccharomyces cerevisiae. Genes Dev *8*, 2349-

REFERENCES

2362.

Kim, J. E., and Chen, J. (2000). Cytoplasmic-nuclear shuttling of FKBP12-rapamycin-associated protein is involved in rapamycin-sensitive signaling and translation initiation. Proc. Natl. Acad. Sci. U.S.A 97, 14340-14345.

Klein, C., and Struhl, K. (1994). Protein kinase A mediates growth-regulated expression of yeast ribosomal protein genes by modulating RAP1 transcriptional activity. Mol. Cell. Biol 14, 1920-1928.

de Kloet, S. R. (1966). Ribonucleic acid synthesis in yeast. The effect of cycloheximide on the synthesis of ribonucleic acid in Saccharomyces carlsbergensis. Biochem. J 99, 566-581.

Knop, M., Siegers, K., Pereira, G., Zachariae, W., Winsor, B., Nasmyth, K., and Schiebel, E. (1999). Epitope tagging of yeast genes using a PCR-based strategy: more tags and improved practical routines. Yeast 15, 963-972.

Kos, M., and Tollervey, D. (2010). Yeast pre-rRNA processing and modification occur cotranscriptionally. Mol. Cell 37, 809-820.

Kressler, D., Linder, P., and de La Cruz, J. (1999). Protein trans-acting factors involved in ribosome biogenesis in Saccharomyces cerevisiae. Mol. Cell. Biol 19, 7897-7912.

Krogan, N. J., Peng, W., Cagney, G., Robinson, M. D., Haw, R., Zhong, G., Guo, X., Zhang, X., Canadien, V., Richards, D. P., et al. (2004). High-definition macromolecular composition of yeast RNA-processing complexes. Mol. Cell 13, 225-239.

Kuhn, C., Geiger, S. R., Baumli, S., Gartmann, M., Gerber, J., Jennebach, S., Mielke, T., Tschochner, H., Beckmann, R., and Cramer, P. (2007). Functional architecture of RNA polymerase I. Cell 131, 1260-1272.

Kulkens, T., Riggs, D. L., Heck, J. D., Planta, R. J., and Nomura, M. (1991). The yeast RNA polymerase I promoter: ribosomal DNA sequences involved in transcription initiation and complex formation in vitro. Nucleic Acids Res 19, 5363-5370.

Laemmli, U. K. (1970). Cleavage of structural proteins during the assembly of the head of bacteriophage T4. Nature 227, 680-685.

Laferté, A., Favry, E., Sentenac, A., Riva, M., Carles, C., and Chédin, S. (2006). The transcriptional activity of RNA polymerase I is a key determinant for the level of all ribosome components. Genes Dev 20, 2030-2040.

Lalo, D., Carles, C., Sentenac, A., and Thuriaux, P. (1993). Interactions between three common subunits of yeast RNA polymerases I and III. Proc. Natl. Acad. Sci. U.S.A 90, 5524-5528.

Lalo, D., Steffan, J. S., Dodd, J. A., and Nomura, M. (1996). RRN11 encodes the third subunit of the complex containing Rrn6p and Rrn7p that is essential for the initiation of rDNA transcription by yeast RNA polymerase I. J. Biol. Chem 271, 21062-21067.

REFERENCES

Lang, W. H., Morrow, B. E., Ju, Q., Warner, J. R., and Reeder, R. H. (1994). A model for transcription termination by RNA polymerase I. Cell 79, 527-534.

Lang, W. H., and Reeder, R. H. (1995). Transcription termination of RNA polymerase I due to a T-rich element interacting with Reb1p. Proc. Natl. Acad. Sci. U.S.A 92, 9781-9785.

Learned, R. M., Cordes, S., and Tjian, R. (1985). Purification and characterization of a transcription factor that confers promoter specificity to human RNA polymerase I. Mol. Cell. Biol 5, 1358-1369.

Léger-Silvestre, I., Trumtel, S., Noaillac-Depeyre, J., and Gas, N. (1999). Functional compartmentalization of the nucleus in the budding yeast Saccharomyces cerevisiae. Chromosoma 108, 103-113.

Léger-Silvestre, I., Caffrey, J. M., Dawaliby, R., Alvarez-Arias, D. A., Gas, N., Bertolone, S. J., Gleizes, P., and Ellis, S. R. (2005). Specific Role for Yeast Homologs of the Diamond Blackfan Anemia-associated Rps19 Protein in Ribosome Synthesis. J. Biol. Chem 280, 38177-38185.

Lempiäinen, H., and Shore, D. (2009). Growth control and ribosome biogenesis. Curr. Opin. Cell Biol 21, 855-863.

Li, H., Tsang, C. K., Watkins, M., Bertram, P. G., and Zheng, X. F. S. (2006). Nutrient regulates Tor1 nuclear localization and association with rDNA promoter. Nature 442, 1058-1061.

Liljelund, P., Mariotte, S., Buhler, J. M., and Sentenac, A. (1992). Characterization and mutagenesis of the gene encoding the A49 subunit of RNA polymerase A in Saccharomyces cerevisiae. Proc. Natl. Acad. Sci. U.S.A 89, 9302-9305.

Lin, C. W., Moorefield, B., Payne, J., Aprikian, P., Mitomo, K., and Reeder, R. H. (1996). A novel 66-kilodalton protein complexes with Rrn6, Rrn7, and TATA-binding protein to promote polymerase I transcription initiation in Saccharomyces cerevisiae. Mol. Cell. Biol 16, 6436-6443.

Lipton, J. M., and Ellis, S. R. (2010). Diamond Blackfan anemia 2008-2009: broadening the scope of ribosome biogenesis disorders. Curr. Opin. Pediatr 22, 12-19.

Lipton, J. M., and Ellis, S. R. (2009). Diamond-Blackfan anemia: diagnosis, treatment, and molecular pathogenesis. Hematol. Oncol. Clin. North Am 23, 261-282.

Loewith, R., and Hall, M. N. (2004). TOR signaling in yeast: temporal and spatial control of cell growth. In Cell growth: control of cell size (Hall, M. N., Raff, M., Thomas, G. Eds.), pp. 139-165. Cold Spring Harbor Laboratory Press, New York.

Loewith, R., Jacinto, E., Wullschleger, S., Lorberg, A., Crespo, J. L., Bonenfant, D., Oppliger, W., Jenoe, P., and Hall, M. N. (2002). Two TOR complexes, only one of which is rapamycin sensitive, have distinct roles in cell growth control. Mol. Cell 10, 457-468.

Lucioli, A., Presutti, C., Ciafrè, S., Caffarelli, E., Fragapane, P., and Bozzoni, I. (1988). Gene dosage alteration of L2 ribosomal protein genes in Saccharomyces cerevisiae: effects on

REFERENCES

ribosome synthesis. Mol. Cell. Biol *8*, 4792-4798.

Marion, R. M., Regev, A., Segal, E., Barash, Y., Koller, D., Friedman, N., and O'Shea, E. K. (2004). Sfp1 is a stress- and nutrient-sensitive regulator of ribosomal protein gene expression. Proc. Natl. Acad. Sci. U.S.A *101*, 14315-14322.

Martin, D. E., and Hall, M. N. (2005). The expanding TOR signaling network. Curr. Opin. Cell Biol *17*, 158-166.

Mason, S. W., Wallisch, M., and Grummt, I. (1997). RNA polymerase I transcription termination: similar mechanisms are employed by yeast and mammals. J. Mol. Biol *268*, 229-234.

Mayer, C., Bierhoff, H., and Grummt, I. (2005). The nucleolus as a stress sensor: JNK2 inactivates the transcription factor TIF-IA and down-regulates rRNA synthesis. Genes Dev *19*, 933-941.

Mayer, C., Zhao, J., Yuan, X., and Grummt, I. (2004). mTOR-dependent activation of the transcription factor TIF-IA links rRNA synthesis to nutrient availability. Genes Dev *18*, 423-434.

Mémet, S., Gouy, M., Marck, C., Sentenac, A., and Buhler, J. M. (1988). RPA190, the gene coding for the largest subunit of yeast RNA polymerase A. J. Biol. Chem *263*, 2830-2839.

Merz, K., Hondele, M., Goetze, H., Gmelch, K., Stoeckl, U., and Griesenbeck, J. (2008). Actively transcribed rRNA genes in S. cerevisiae are organized in a specialized chromatin associated with the high-mobility group protein Hmo1 and are largely devoid of histone molecules. Genes Dev *22*, 1190-1204.

Milkereit, P., Schultz, P., and Tschochner, H. (1997). Resolution of RNA polymerase I into dimers and monomers and their function in transcription. Biol. Chem *378*, 1433-1443.

Milkereit, P., and Tschochner, H. (1998). A specialized form of RNA polymerase I, essential for initiation and growth-dependent regulation of rRNA synthesis, is disrupted during transcription. EMBO J *17*, 3692-3703.

Miller, G., Panov, K. I., Friedrich, J. K., Trinkle-Mulcahy, L., Lamond, A. I., and Zomerdijk, J. C. (2001). hRRN3 is essential in the SL1-mediated recruitment of RNA Polymerase I to rRNA gene promoters. EMBO J *20*, 1373-1382.

Miller, O. L., and Beatty, B. R. (1969). Visualization of nucleolar genes. Science *164*, 955-957.

Mitsui, K., Nakagawa, T., and Tsurugi, K. (1988). On the size and the role of a free cytosolic pool of acidic ribosomal proteins in yeast Saccharomyces cerevisiae. J. Biochem *104*, 908-911.

Moorefield, B., Greene, E. A., and Reeder, R. H. (2000). RNA polymerase I transcription factor Rrn3 is functionally conserved between yeast and human. Proc. Natl. Acad. Sci. U.S.A *97*, 4724-4729.

Moss, T., Langlois, F., Gagnon-Kugler, T., and Stefanovsky, V. (2007). A housekeeper with power of

REFERENCES

attorney: the rRNA genes in ribosome biogenesis. Cell. Mol. Life Sci *64*, 29-49.

Moss, T. (2004). At the crossroads of growth control; making ribosomal RNA. Curr. Opin. Genet. Dev *14*, 210-217.

Muratani, M., and Tansey, W. P. (2003). How the ubiquitin-proteasome system controls transcription. Nat. Rev. Mol. Cell Biol *4*, 192-201.

Musters, W., Knol, J., Maas, P., Dekker, A. F., van Heerikhuizen, H., and Planta, R. J. (1989). Linker scanning of the yeast RNA polymerase I promoter. Nucleic Acids Res *17*, 9661-9678.

Nadeem, F. K., Blair, D., and McInerny, C. J. (2006). Pol5p, a novel binding partner to Cdc10p in fission yeast involved in rRNA production. Mol. Genet. Genomics *276*, 391-401.

Nogi, Y., Yano, R., Dodd, J., Carles, C., and Nomura, M. (1993). Gene RRN4 in Saccharomyces cerevisiae encodes the A12.2 subunit of RNA polymerase I and is essential only at high temperatures. Mol. Cell. Biol *13*, 114-122.

Oakes, M. L., Siddiqi, I., French, S. L., Vu, L., Sato, M., Aris, J. P., Beyer, A. L., and Nomura, M. (2006). Role of histone deacetylase Rpd3 in regulating rRNA gene transcription and nucleolar structure in yeast. Mol. Cell. Biol *26*, 3889-3901.

Oeffinger, M., Dlakic, M., and Tollervey, D. (2004). A pre-ribosome-associated HEAT-repeat protein is required for export of both ribosomal subunits. Genes Dev *18*, 196-209.

Osheim, Y. N., French, S. L., Keck, K. M., Champion, E. A., Spasov, K., Dragon, F., Baserga, S. J., and Beyer, A. L. (2004). Pre-18S ribosomal RNA is structurally compacted into the SSU processome prior to being cleaved from nascent transcripts in Saccharomyces cerevisiae. Mol. Cell *16*, 943-954.

Panov, K. I., Friedrich, J. K., Russell, J., and Zomerdijk, J. C. B. M. (2006a). UBF activates RNA polymerase I transcription by stimulating promoter escape. EMBO J *25*, 3310-3322.

Panov, K. I., Panova, T. B., Gadal, O., Nishiyama, K., Saito, T., Russell, J., and Zomerdijk, J. C. B. M. (2006b). RNA polymerase I-specific subunit CAST/hPAF49 has a role in the activation of transcription by upstream binding factor. Mol. Cell. Biol *26*, 5436-5448.

Paule, M. R. (1998). Promoter structure of class I genes. In Transcription of ribosomal RNA genes by eukaryotic RNA polymerase I (Paule, M. R. Ed.), pp. 39-50. Landes Bioscience, Austin.

Paule, M. R., Iida, C. T., Perna, P. J., Harris, G. H., Knoll, D. A., and D'Alessio, J. M. (1984). In vitro evidence that eukaryotic ribosomal RNA transcription is regulated by modification of RNA polymerase I. Nucleic Acids Res *12*, 8161-8180.

Peng, W. T., Robinson, M. D., Mnaimneh, S., Krogan, N. J., Cagney, G., Morris, Q., Davierwala, A. P., Grigull, J., Yang, X., Zhang, W., et al. (2003). A panoramic view of yeast noncoding RNA processing. Cell *113*, 919-933.

Petes, T. D. (1979). Yeast ribosomal DNA genes are located on chromosome XII. Proc. Natl. Acad.

REFERENCES

Sci. U.S.A 76, 410-414.

Peyroche, G., Milkereit, P., Bischler, N., Tschochner, H., Schultz, P., Sentenac, A., Carles, C., and Riva, M. (2000). The recruitment of RNA polymerase I on rDNA is mediated by the interaction of the A43 subunit with Rrn3. EMBO J 19, 5473-5482.

Peyroche, G., Levillain, E., Siaut, M., Callebaut, I., Schultz, P., Sentenac, A., Riva, M., and Carles, C. (2002). The A14-A43 heterodimer subunit in yeast RNA pol I and their relationship to Rpb4-Rpb7 pol II subunits. Proc. Natl. Acad. Sci. U.S.A 99, 14670-14675.

Philippi, A. (2008). Untersuchungen zur Rolle des Transkriptionsfaktors Rrn3p in der wachstumsabhängigen Regulation der rRNA-Synthese.

Philippi, A., Steinbauer, R., Reiter, A., Fath, S., Leger-Silvestre, I., Milkereit, P., Griesenbeck, J., and Tschochner, H. (2010). TOR-dependent reduction in the expression level of Rrn3p lowers the activity of the yeast RNA Pol I machinery, but does not account for the strong inhibition of rRNA production. Nucleic Acids Res. Available at: http://www.ncbi.nlm.nih.gov/pubmed/ 20421203 [Accessed August 9, 2010].

Planta, R. J., and Mager, W. H. (1998). The list of cytoplasmic ribosomal proteins of Saccharomyces cerevisiae. Yeast 14, 471-477.

Pleiss, J. A., Whitworth, G. B., Bergkessel, M., and Guthrie, C. (2007). Rapid, transcript-specific changes in splicing in response to environmental stress. Mol. Cell 27, 928-937.

Pöll, G., Braun, T., Jakovljevic, J., Neueder, A., Jakob, S., Woolford, J. L., Tschochner, H., and Milkereit, P. (2009). rRNA maturation in yeast cells depleted of large ribosomal subunit proteins. PLoS ONE 4, e8249.

Powers, T., and Walter, P. (1999). Regulation of ribosome biogenesis by the rapamycin-sensitive TOR-signaling pathway in Saccharomyces cerevisiae. Mol. Biol. Cell 10, 987-1000.

Prescott, E. M., Osheim, Y. N., Jones, H. S., Alen, C. M., Roan, J. G., Reeder, R. H., Beyer, A. L., and Proudfoot, N. J. (2004). Transcriptional termination by RNA polymerase I requires the small subunit Rpa12p. Proc. Natl. Acad. Sci. U.S.A 101, 6068-6073.

Prieto, J., and McStay, B. (2007). Recruitment of factors linking transcription and processing of pre-rRNA to NOR chromatin is UBF-dependent and occurs independent of transcription in human cells. Genes Dev 21, 2041-2054.

Puig, O., Caspary, F., Rigaut, G., Rutz, B., Bouveret, E., Bragado-Nilsson, E., Wilm, M., and Séraphin, B. (2001). The tandem affinity purification (TAP) method: a general procedure of protein complex purification. Methods 24, 218-229.

Reeder, R. H., and Lang, W. H. (1998). Stopping RNA polymerase I. In Transcription of ribosomal RNA genes by eukaryotic RNA polymerase I (Paule, M. R. Ed.), pp. 173-178. Landes Bioscience, Austin.

Robledo, S., Idol, R. A., Crimmins, D. L., Ladenson, J. H., Mason, P. J., and Bessler, M. (2008). The

REFERENCES

role of human ribosomal proteins in the maturation of rRNA and ribosome production. RNA *14*, 1918-1929.

Rosbash, M., Harris, P. K., Woolford, J. L., and Teem, J. L. (1981). The effect of temperature-sensitive RNA mutants on the transcription products from cloned ribosomal protein genes of yeast. Cell *24*, 679-686.

Rubin, D. M., Coux, O., Wefes, I., Hengartner, C., Young, R. A., Goldberg, A. L., and Finley, D. (1996). Identification of the gal4 suppressor Sug1 as a subunit of the yeast 26S proteasome. Nature *379*, 655-657.

Rudra, D., and Warner, J. R. (2004). What better measure than ribosome synthesis? Genes Dev *18*, 2431-2436.

Rudra, D., Zhao, Y., and Warner, J. R. (2005). Central role of Ifh1p-Fhl1p interaction in the synthesis of yeast ribosomal proteins. EMBO J *24*, 533-542.

Russell, J., and Zomerdijk, J. C. B. M. (2005). RNA-polymerase-I-directed rDNA transcription, life and works. Trends Biochem. Sci *30*, 87-96.

Schawalder, S. B., Kabani, M., Howald, I., Choudhury, U., Werner, M., and Shore, D. (2004). Growth-regulated recruitment of the essential yeast ribosomal protein gene activator Ifh1. Nature *432*, 1058-1061.

Schlosser, A., Bodem, J., Bossemeyer, D., Grummt, I., and Lehmann, W. D. (2002). Identification of protein phosphorylation sites by combination of elastase digestion, immobilized metal affinity chromatography, and quadrupole-time of flight tandem mass spectrometry. Proteomics *2*, 911-918.

Schmid, M., Durussel, T., and Laemmli, U. K. (2004). ChIC and ChEC; genomic mapping of chromatin proteins. Mol. Cell *16*, 147-157.

Schmitt, M. E., Brown, T. A., and Trumpower, B. L. (1990). A rapid and simple method for preparation of RNA from Saccharomyces cerevisiae. Nucleic Acids Res *18*, 3091-3092.

Schmitt, M. E., and Clayton, D. A. (1993). Nuclear RNase MRP is required for correct processing of pre-5.8S rRNA in Saccharomyces cerevisiae. Mol. Cell. Biol *13*, 7935-7941.

Schnapp, A., Schnapp, G., Erny, B., and Grummt, I. (1993). Function of the growth-regulated transcription initiation factor TIF-IA in initiation complex formation at the murine ribosomal gene promoter. Mol. Cell. Biol *13*, 6723-6732.

Schnapp, G., Graveley, B. R., and Grummt, I. (1996). TFIIS binds to mouse RNA polymerase I and stimulates transcript elongation and hydrolytic cleavage of nascent rRNA. Mol. Gen. Genet *252*, 412-419.

Schneider, D. A., French, S. L., Osheim, Y. N., Bailey, A. O., Vu, L., Dodd, J., Yates, J. R., Beyer, A. L., and Nomura, M. (2006). RNA polymerase II elongation factors Spt4p and Spt5p play roles in transcription elongation by RNA polymerase I and rRNA processing. Proc. Natl. Acad.

REFERENCES

Sci. U.S.A *103*, 12707-12712.

Schneider, D. A., Michel, A., Sikes, M. L., Vu, L., Dodd, J. A., Salgia, S., Osheim, Y. N., Beyer, A. L., and Nomura, M. (2007). Transcription elongation by RNA polymerase I is linked to efficient rRNA processing and ribosome assembly. Mol. Cell *26*, 217-229.

Schweizer, E., MacKechnie, C., and Halvorson, H. O. (1969). The redundancy of ribosomal and transfer RNA genes in Saccharomyces cerevisiae. J. Mol. Biol *40*, 261-277.

Sehgal, S. N., Baker, H., and Vézina, C. (1975). Rapamycin (AY-22,989), a new antifungal antibiotic. II. Fermentation, isolation and characterization. J. Antibiot *28*, 727-732.

Sherman, F. (2002). Getting started with yeast. Meth. Enzymol *350*, 3-41.

Shevchenko, A., Wilm, M., Vorm, O., and Mann, M. (1996). Mass spectrometric sequencing of proteins silver-stained polyacrylamide gels. Anal. Chem *68*, 850-858.

Shevchenko, A., Tomas, H., Havlis, J., Olsen, J. V., and Mann, M. (2006). In-gel digestion for mass spectrometric characterization of proteins and proteomes. Nat Protoc *1*, 2856-2860.

Shimizu, K., Kawasaki, Y., Hiraga, S., Tawaramoto, M., Nakashima, N., and Sugino, A. (2002). The fifth essential DNA polymerase phi in Saccharomyces cerevisiae is localized to the nucleolus and plays an important role in synthesis of rRNA. Proc. Natl. Acad. Sci. U.S.A *99*, 9133-9138.

Shou, W., Sakamoto, K. M., Keener, J., Morimoto, K. W., Traverso, E. E., Azzam, R., Hoppe, G. J., Feldman, R. M., DeModena, J., Moazed, D., et al. (2001). Net1 stimulates RNA polymerase I transcription and regulates nucleolar structure independently of controlling mitotic exit. Mol. Cell *8*, 45-55.

Sikorski, R. S., and Hieter, P. (1989). A system of shuttle vectors and yeast host strains designed for efficient manipulation of DNA in Saccharomyces cerevisiae. Genetics *122*, 19-27.

Smid, A., Riva, M., Bouet, F., Sentenac, A., and Carles, C. (1995). The association of three subunits with yeast RNA polymerase is stabilized by A14. J. Biol. Chem *270*, 13534-13540.

Song, J. M., Cheung, E., and Rabinowitz, J. C. (1996). Organization and characterization of the two yeast ribosomal protein YL19 genes. Curr. Genet *30*, 273-278.

Soulard, A., Cohen, A., and Hall, M. N. (2009). TOR signaling in invertebrates. Curr. Opin. Cell Biol *21*, 825-836.

Spingola, M., Grate, L., Haussler, D., and Ares, M. (1999). Genome-wide bioinformatic and molecular analysis of introns in Saccharomyces cerevisiae. RNA *5*, 221-234.

Stefanovsky, V., Langlois, F., Gagnon-Kugler, T., Rothblum, L. I., and Moss, T. (2006). Growth factor signaling regulates elongation of RNA polymerase I transcription in mammals via UBF phosphorylation and r-chromatin remodeling. Mol. Cell *21*, 629-639.

REFERENCES

Steffan, J. S., Keys, D. A., Dodd, J. A., and Nomura, M. (1996). The role of TBP in rDNA transcription by RNA polymerase I in Saccharomyces cerevisiae: TBP is required for upstream activation factor-dependent recruitment of core factor. Genes Dev *10*, 2551-2563.

Steffan, J. S., Keys, D. A., Vu, L., and Nomura, M. (1998). Interaction of TATA-binding protein with upstream activation factor is required for activated transcription of ribosomal DNA by RNA polymerase I in Saccharomyces cerevisiae in vivo. Mol. Cell. Biol *18*, 3752-3761.

Steinbauer, R. (2006). Posttranslationale Modifikationen und Funktion des Transkriptionsfaktors Rrn3p.

Tavner, F. J., Simpson, R., Tashiro, S., Favier, D., Jenkins, N. A., Gilbert, D. J., Copeland, N. G., Macmillan, E. M., Lutwyche, J., Keough, R. A., et al. (1998). Molecular cloning reveals that the p160 Myb-binding protein is a novel, predominantly nucleolar protein which may play a role in transactivation by Myb. Mol. Cell. Biol *18*, 989-1002.

Thomas, B. J., and Rothstein, R. (1989). Elevated recombination rates in transcriptionally active DNA. Cell *56*, 619-630.

Thomas, G., and Hall, M. N. (1997). TOR signalling and control of cell growth. Curr. Opin. Cell Biol *9*, 782-787.

Tower, J., and Sollner-Webb, B. (1987). Transcription of mouse rDNA is regulated by an activated subform of RNA polymerase I. Cell *50*, 873-883.

Tschochne, H., and Milkereit, P. (1997). RNA polymerase I from S. cerevisiae depends on an additional factor to release terminated transcripts from the template. FEBS Lett *410*, 461-466.

Tschochner, H. (1996). A novel RNA polymerase I-dependent RNase activity that shortens nascent transcripts from the 3' end. Proc. Natl. Acad. Sci. U.S.A *93*, 12914-12919.

Tschochner, H., and Hurt, E. (2003). Pre-ribosomes on the road from the nucleolus to the cytoplasm. Trends Cell Biol *13*, 255-263.

Udem, S. A., and Warner, J. R. (1972). Ribosomal RNA synthesis in Saccharomyces cerevisiae. J. Mol. Biol *65*, 227-242.

Urban, J., Soulard, A., Huber, A., Lippman, S., Mukhopadhyay, D., Deloche, O., Wanke, V., Anrather, D., Ammerer, G., Riezman, H., et al. (2007). Sch9 is a major target of TORC1 in Saccharomyces cerevisiae. Mol. Cell *26*, 663-674.

Vanrobays, E., Leplus, A., Osheim, Y. N., Beyer, A. L., Wacheul, L., and Lafontaine, D. L. J. (2008). TOR regulates the subcellular distribution of DIM2, a KH domain protein required for cotranscriptional ribosome assembly and pre-40S ribosome export. RNA *14*, 2061-2073.

Venema, J., and Tollervey, D. (1999). Ribosome synthesis in Saccharomyces cerevisiae. Annu. Rev. Genet *33*, 261-311.

REFERENCES

Vézina, C., Kudelski, A., and Sehgal, S. N. (1975). Rapamycin (AY-22,989), a new antifungal antibiotic. I. Taxonomy of the producing streptomycete and isolation of the active principle. J. Antibiot 28, 721-726.

Wai, H., Johzuka, K., Vu, L., Eliason, K., Kobayashi, T., Horiuchi, T., and Nomura, M. (2001). Yeast RNA polymerase I enhancer is dispensable for transcription of the chromosomal rRNA gene and cell growth, and its apparent transcription enhancement from ectopic promoters requires Fob1 protein. Mol. Cell. Biol 21, 5541-5553.

Warner, J. R. (1977). In the absence of ribosomal RNA synthesis, the ribosomal proteins of HeLa cells are synthesized normally and degraded rapidly. J. Mol. Biol 115, 315-333.

Warner, J. R. (1989). Synthesis of ribosomes in Saccharomyces cerevisiae. Microbiol. Rev 53, 256-271.

Warner, J. R. (1999). The economics of ribosome biosynthesis in yeast. Trends Biochem. Sci 24, 437-440.

Warner, J. R., Mitra, G., Schwindinger, W. F., Studeny, M., and Fried, H. M. (1985). Saccharomyces cerevisiae coordinates accumulation of yeast ribosomal proteins by modulating mRNA splicing, translational initiation, and protein turnover. Mol. Cell. Biol 5, 1512-1521.

Warner, J. R., and Udem, S. A. (1972). Temperature sensitive mutations affecting ribosome synthesis in Saccharomyces cerevisiae. J. Mol. Biol 65, 243-257.

Wei, Y., Tsang, C. K., and Zheng, X. F. S. (2009). Mechanisms of regulation of RNA polymerase III-dependent transcription by TORC1. EMBO J 28, 2220-2230.

Wery, M., Ruidant, S., Schillewaert, S., Leporé, N., and Lafontaine, D. L. J. (2009). The nuclear poly(A) polymerase and Exosome cofactor Trf5 is recruited cotranscriptionally to nucleolar surveillance. RNA 15, 406-419.

Wessel, D., and Flügge, U. I. (1984). A method for the quantitative recovery of protein in dilute solution in the presence of detergents and lipids. Anal. Biochem 138, 141-143.

Wilson, D. N., and Nierhaus, K. H. (2003). The ribosome through the looking glass. Angew. Chem. Int. Ed. Engl 42, 3464-3486.

Wittekind, M., Kolb, J. M., Dodd, J., Yamagishi, M., Mémet, S., Buhler, J. M., and Nomura, M. (1990). Conditional expression of RPA190, the gene encoding the largest subunit of yeast RNA polymerase I: effects of decreased rRNA synthesis on ribosomal protein synthesis. Mol. Cell. Biol 10, 2049-2059.

Wullschleger, S., Loewith, R., and Hall, M. N. (2006). TOR signaling in growth and metabolism. Cell 124, 471-484.

Wullschleger, S., Loewith, R., Oppliger, W., and Hall, M. N. (2005). Molecular organization of target of rapamycin complex 2. J. Biol. Chem 280, 30697-30704.

REFERENCES

Yamamoto, R. T., Nogi, Y., Dodd, J. A., and Nomura, M. (1996). RRN3 gene of Saccharomyces cerevisiae encodes an essential RNA polymerase I transcription factor which interacts with the polymerase independently of DNA template. EMBO J 15, 3964-3973.

Yang, W., Rogozin, I. B., and Koonin, E. V. (2003). Yeast POL5 is an evolutionarily conserved regulator of rDNA transcription unrelated to any known DNA polymerases. Cell Cycle 2, 120-122.

Yuan, X., Zhao, J., Zentgraf, H., Hoffmann-Rohrer, U., and Grummt, I. (2002). Multiple interactions between RNA polymerase I, TIF-IA and TAF(I) subunits regulate preinitiation complex assembly at the ribosomal gene promoter. EMBO Rep 3, 1082-1087.

Zaragoza, D., Ghavidel, A., Heitman, J., and Schultz, M. C. (1998). Rapamycin induces the G0 program of transcriptional repression in yeast by interfering with the TOR signaling pathway. Mol. Cell. Biol 18, 4463-4470.

Zhang, Y., Sikes, M. L., Beyer, A. L., and Schneider, D. A. (2009). The Paf1 complex is required for efficient transcription elongation by RNA polymerase I. Proc. Natl. Acad. Sci. U.S.A 106, 2153-2158.

Zhang, Y., Smith, A. D., Renfrow, M. B., and Schneider, D. A. (2010). The RNA polymerase-associated factor 1 complex (Paf1C) directly increases the elongation rate of RNA polymerase I and is required for efficient regulation of rRNA synthesis. J. Biol. Chem 285, 14152-14159.

Zhao, J., Yuan, X., Frödin, M., and Grummt, I. (2003). ERK-dependent phosphorylation of the transcription initiation factor TIF-IA is required for RNA polymerase I transcription and cell growth. Mol. Cell 11, 405-413.

REFERENCES

7 PUBLICATIONS

Reiter*, A., Steinbauer*, R., Philippi*, A., Gerber, J., Tschochner, H., Milkereit, P., Griesenbeck, J. (2010). Reduction in ribosomal protein synthesis is sufficient to explain major effects on ribosome production after short-term TOR inactivation in Saccharomyces cerevisiae. Mol. Cell. Biol (accepted)

Philippi*, A., Steinbauer*, R., Reiter*, A., Fath, S., Leger-Silvestre, I., Milkereit, P., Griesenbeck, J., Tschochner, H. (2010). TOR-dependent reduction in the expression level of Rrn3p lowers the activity of the yeast RNA Pol I machinery, but does not account for the strong inhibition of rRNA production. Nucleic Acids Research 38, 5315-5326

Gerber, J., Reiter, A., Steinbauer, R., Jakob, S., Kuhn, C., Cramer, P., Griesenbeck, J., Milkereit, P., Tschochner, H. (2008). Site specific phosphorylation of yeast RNA polymerase I. Nucleic Acids Research 36, 793-802

* contributed equally

PUBLICATIONS

8 ABBREVIATIONS

A	ampere
amp	ampicillin
A/C	autonomous replication sequence/centromere (single copy)
ATP	adenosin triphosphate
bp	base pair(s)
CHCA	α-cyano-4-hydroxycinnamic acid
CSM	complete supplement mixture
C-terminal	carboxy-terminal
Da	dalton
DIC	differential contrast
DNA	desoxyribonucleic acid
dNTP	desoxyribonucleoside-5'-triphosphate
ETS	external transcribed spacer
g	gram(s)
gen	geneticin
GFP	green fluorescence protein
GST	glutathione S-transferase
h	hour(s)
HA_3	triple hemagglutinin
hyg	hygromycin B
IN	input
IP	immunoprecipitation
ITS	internal transcribed spacer
k	kilo(s)
kb	kilo base pair(s)
l	liter(s)
LSU	large ribosomal subunit
m	milli / meter
M	molar (mol/l)
MALDI	matrix assisted laser desorption/ionization
min	minute(s)
mRNA	messenger RNA
MS	mass spectrometry
MS/MS	tandem mass spectrometry
n	nano
NTS	non-transcribed spacer

ABBREVIATIONS

OD_{600}	optical density at 600 nm
PAF	polymerase associated factor
H	negative decadic logarithm [H$^+$]
PMF	peptide mass fingerprint
PMSF	phenylmethylsulfonyl fluoride
Prot.A	protein A
rDNA	ribosomal DNA
RNA	ribonucleic acid
rpm	rotations per minute
S	sedimentation coefficient
s/sec	second(s)
Taq	thermus aquaticus
TAP	tandem affinity tag
TOF	time of flight
U	unit(s)
V	volt
v/v	volume/volume
WT	wild type
w/v	weight/volume
YNB	yeast nitrogen base
µ	micro
2µ	2 micron (multi copy)

Acknowledgments

Abschließend möchte ich mich ganz herzlich bei allen bedanken, die zum Gelingen dieser Arbeit beigetragen haben:

In erster Linie gilt mein Dank natürlich Prof. Dr. Herbert Tschochner, der mir die Möglichkeit gegeben hat in seinem Labor an diesem interessanten Thema zu arbeiten und für die damit verbundene tolle Betreuung während dieser Zeit.

Ganz besonders bedanken möchte ich mich sowohl bei Dr. Joachim Griesenbeck als auch bei Dr. Philipp Milkereit für die vielen produktiven Anregungen und ihre stete Bereitschaft über die Ergebnisse meiner Arbeit zu diskutieren.

Ich möchte mich ferner bei meinen Kollegen der Subgruppe „Transkription" für die gute Zusammenarbeit bedanken, insbesondere bei Alarich Reiter, unter anderem für die Erlaubnis eines seiner Resultate abbilden zu dürfen, und bei Dr. Jochen Gerber, dessen praktische Ratschläge mir im Laboralltag oft geholfen haben.

Recht herzlich bedanken möchte ich mich bei allen Mitgliedern des „House of the Ribosome" für ihre Hilfsbereitschaft und nicht zuletzt für das sehr angenehme Arbeitsklima.

Meinen Eltern, meinen Geschwistern und meiner Oma, die leider den Abschluss dieser Arbeit nicht mehr miterleben konnte, danke ich für ihre unermüdliche moralische Unterstützung und meinen Eltern insbesondere dafür, dass sie mir diese Ausbildung überhaupt erst ermöglicht haben.

Schließlich möchte ich mich bei meiner Freundin Anja bedanken, die großen Anteil am Gelingen dieser Arbeit hat, da sie mich während der ganzen Zeit auf jede erdenkliche Weise unterstützt hat und immer für mich da war.

I want morebooks!

Buy your books fast and straightforward online - at one of world's fastest growing online book stores! Environmentally sound due to Print-on-Demand technologies.

Buy your books online at
www.morebooks.shop

Kaufen Sie Ihre Bücher schnell und unkompliziert online – auf einer der am schnellsten wachsenden Buchhandelsplattformen weltweit! Dank Print-On-Demand umwelt- und ressourcenschonend produziert.

Bücher schneller online kaufen
www.morebooks.shop

KS OmniScriptum Publishing
Brivibas gatve 197
LV-1039 Riga, Latvia
Telefax: +371 686 204 55

info@omniscriptum.com
www.omniscriptum.com

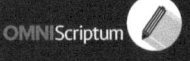

Printed by Books on Demand GmbH, Norderstedt / Germany